Content

Complexity

Information Design
in Technical Communication

Content

Complexity

Information Design
in Technical Communication

Edited by

Michael J. Albers
University of Memphis

Beth Mazur
University of Baltimore

LEA   LAWRENCE ERLBAUM ASSOCIATES, PUBLISHERS
2003   Mahwah, New Jersey   London

Library
University of Texas
at San Antonio

Lawrence Erlbaum Associates, Inc., Publishers
10 Industrial Avenue
Mahwah, NJ 07430

Cover design by Kathryn Houghtaling Lacey

Content & complexity : information design in technical communication / edited by Michael J. Albers, Beth Mazur.
    p. cm.
    Includes bibliographical references and index.
ISBN 0-8058-4140-7 (c : alk. paper)
ISBN 0-8058-4141-5 (pbk.)
1. Communication of technical information. 2. System design. 3. Visual communication. 4. Written communication. I. Albers, Michael J. II. Mazur, Beth.
T10.5 .C65 2002
005.1′5—dc21   2002192714

Books published by Lawrence Erlbaum Associates are printed on acid-free paper, and their bindings are chosen for strength and durability.

Printed in the United States of America
10 9 8 7 6 5 4 3 2 1

Acknowledgments

Any anthology is obviously the work of many people who are well deserving of thanks.

First there are our Erlbaum editors. Linda Bathgate, our acquisition editor, helped us throughout the proposal submission process. Sondra Guideman, our production editor, helped throughout the production process. Thank you for your help! We'd also like to thank the proposal reviewers that provided invaluable feedback on shaping the final focus of the book.

Thank you especially to the authors who wrote the chapters of this book—and journeyed with us through the stages from concept to finished book. Without them it would be a very thin volume!

We'd like to offer a special thanks to Karen Schriver for reviewing the proposal and writing the foreword. Finally, many, many thanks to the STC Information Design Special Interest Group, whose members provided us with the original idea for this volume and ongoing support in our quest to explore information design in the context of technical communication.

Michael J. Albers

Beth Mazur

Contents

Foreword

Karen Schriver
KSA Document Design & Research, Inc.

Information designers bring together words and images in ways that enable people to understand, take action, or make decisions. A good information design helps people to use the content in ways that suit their unique interests. Although information designers have long recognized the importance of developing good content, much of the literature in the field has focused on issues of graphic design and typography. This volume broadens our perspective with new ideas about creating information designs that speak to peoples' needs through the design of effective content in products familiar to technical communicators.

Since the 1990s, information designers have been preoccupied with shaping content in order to reduce information overload. The authors here challenge us to think strategically about content—its selection, organization, and integration. They show that well designed content can help people ferret out fact from fiction, main points from details, and "must read" from "optional read" information. In addition, the authors remind us that the content we generate not only communicates information to people but helps to build relationships among people.

Throughout the book, a number of themes emerge. Perhaps the most prominent is the need to hone our skills in analyzing the structure of information more deliberately than we have in the past. For any given information design task—whether a paper document, an online help system, or a multimedia project—information designers need to identify core (must have) information and distinguish this content from supplemental (nice to have) information. With a thorough understanding of the structure, information designers can highlight the content distinctions through careful design of text, graphics, photography, full-motion video, typography, or sound.

Although making these sorts of structural distinctions apparent for readers has always proved challenging, the difficulties have increased as the media and technologies for presenting information have evolved. The authors here argue that information designers need to sharpen their talents in making these distinctions visible, whatever technologies or media they are using. For example, the design of information must enable readers to distinguish between core and supplemental information whether the message is delivered as any of the following:

- A paper artifact
- An online document or database (e.g., on the Web or on a CD-ROM)
- A website with links to related information
- A document presented in telegraphic chunks on a very small screen (e.g., a watch, a cellular phone, or a Palm Pilot)
- An online environment using technologies that intelligently and dynamically adapt the content presented
- A document-like artifact generated dynamically on-the-fly from a single-source database of online information

Because different media place different constraints on the amount of information one can display, media differences pose significant challenges both for designing good content and for making obvious what content is available.

A related theme that emerges in this book is that information designers need to be more critical about the nature of the content they present. Simply attending to whether the content is clear is not enough. The authors point out that too often readers are presented with simple, clear, yet inappropriate information. For example, a user of a presentation software program might call on a help system to tell her how to put page numbers on a handout for members of her audience. If she searches the help files and finds only information on paginating slides rather than handouts, that content is useless even if well formed. Similarly, if a bank customer checks his monthly statement for the interest rate on cash advances and finds instead information about interest rates for a credit card insurance plan, that information is useless.

The authors tell us that familiar phrases such as "easy-to-use" and "easy-to-access" can be merely slogans if the content is inappropriate. Information that can be retrieved in a few seconds and that looks "short and snappy" can be deemed useless if the content is ambiguous, abstract, leaves out critical detail, or is simply wrong. Unfortunately, many organizations select and organize their content by default, that is, according to their own development processes—creating content not by design, but by historical accident and accretion. Without detailed information about their stakeholders' expectations and needs for content, organizations can produce artifacts that fail, even though they look nice and read well. Work presented in this book tells us that we need to rethink our ideas about content and about whose needs should take priority.

Of course, there are a variety of stakeholders for most information designs. The authors here suggest that information designers refine their methods for understanding the diversity among stakeholders—their topic knowledge, experience, needs, wants, and motivations. Just whose needs should take priority is not a simple matter. Although some stakeholders will be inexperienced with the subject matter and will seek introductory content about the topic, other people experienced in the subject matter will be frustrated with a mere introduction. An approach to designing content that is appropriate for someone with little knowledge may be quite inadequate for someone with intermediate or expert knowledge of the topic.

Most of the work in information design has not yet explored the needs of people who possess high knowledge or experience in a domain. In fact, many information designs seem to ignore stakeholders with complex questions and ill-structured problems. People who seek answers to complex questions often need to examine a sequence of information artifacts en route to finding a solution. They may search for relevant content across textual material, maps, tables, photographs, or arrays of numerical data. For instance, a physician trying to understand the data from clinical trials for competitive cancer drugs might consult journal articles and websites in order to compare the data those sources offer with what he may receive in the marketing literature from pharmaceutical companies.

How best to orchestrate content across information artifacts to facilitate high-level problem solving has rarely been studied. Researchers need to address this issue, paying special attention to the ways that people represent their problems when they turn to information designs. With better knowledge of their stakeholders' worlds, information designers will be better positioned to imagine artifacts that will be more responsive, more useful. The authors here argue that information designers should take a closer look at the world of their stakeholders.

Some information designers would argue that they have always helped stakeholders solve difficult problems, and in doing so, have made a career out of making the complex clear. Yet often missing from the information designer's repertoire has been the ability to create designs that enable people to embrace complexity and use it to their advantage. Learning to make the complex both visible and useful represents a significant challenge for the field.

Equally important, information designers need to step back from the glow of the screen to see that they are not so much designing stand-alone information products as much as they are creating artifacts that enable relationships among people. Information designers ought to be concerned with the things that go on in the everyday world of people engaging with communications at bank counters and help desks, in cubicles and checkout counters, in the living room or the classroom. Because information design is a kind of conversation, it is important that members of the field think deeply about their writing and design. They must take up the challenge of setting the tone for the conversation and of deciding what to say and how to say it.

Information designers have had an abiding interest in creating content that is helpful, useful, and truthful. Indeed, the stakes for good content have always been high. Even so, our ideas about content have been neither fully developed nor fully elaborated. The authors of this volume extend what we know about shaping content, both visually and verbally. They help us to see how the design of content influences both peoples' understanding of the subject matter and their understanding of those doing the talking.

Introduction

Michael J. Albers
University of Memphis

Getting a grip on information complexity

High-quality information design communicates information in a manner appropriate and pertinent to a reader's situational context. It must focus on the reader and ensure that he or she can clearly extract the information needed to accomplish the real-world goal which sent them searching for information.

In its general sense, information design ranges from developing maps and signage to web pages (Jacobson, 1999). Although in no way attempting to limit the scope of information design, this book works with information as it applies in technical communication, particularly as practiced within the software industry.

As a discipline, information design has only recently gained visibility. It has emerged from a melting of various fields, primarily graphical design, human factors, and technical communication. (This melting is discussed later in this chapter.) The need for information design arises because of the increasing complexity and volume of information that people are expected to process. Tufte (1983) discussed this problem in his first book, *The Visual Display of Quantitative Information*, and, if anything, the problem has gotten worse with the deluge of information available via the World Wide Web or corporate intranets. People simply cannot efficiently sort through and process the amount of information they have access to; information overload has become a major problem. To reach the answer, they need content properly positioned within the problem's context and effectively assembled and presented.

A definition of information design

It is ironic that defining information design is a major problem facing the information design discipline. Any article or book on information design will have a definition in it, ranging from being synonymous with graphic design to web design to technical communication, and all variations between. The definitions never match. In a special issue of *Technical Communication* (May, 2000), a section on information design contained at least three definitions in the introduction and commentaries alone.

But what is information design? The Vienna-based International Institute of Information Design (2000) admits that information design

> can be hard to define, because it is an interdisciplinary approach which combines skills in graphic design, writing and editing, illustration, and human factors. Information designers seek to combine skills in these fields to make complex information easier to understand.

Before considering various definitions of information design, I must solidly reject some definitions I have found. Actually, no one has explicitly defined information design in these terms, rather the definition was apparent from context of their usage. Information design is not equivalent to or a synonym for:

- Graphical design
- Web design
- Information architecture

Various leading figures—who come from a language-based background in information design—have placed their own definitions on the table. But as you can see reading these definitions, they come in radically different flavors, although they contain similar elements. (For more definitions of information design, see Carliner, in this volume.)

Janice Redish

Information design is what we do to develop a document (or communication) that works for its users. Working for its users means that the people who must or want to use the information can:

- find what they need,
- understand what they find, and
- use what they understand appropriately.

Conrad Taylor

One may be a professional writer or illustrator or designer, or combine these into the profession of technical communicator; but I (somewhat controversially) don't regard information design as a profession. I see it more as a stance that one takes, like a political or moral stance. Whether we create software manuals or street signs or schoolbooks, by aligning ourselves with the aspirations of information design, we are making a promise to ourselves and each other to improve the quality of communication, respecting and improving the lives of our fellows.

Information Design Journal

Information design is the art and the science of presenting information so that it is understandable and easy to use: effective, efficient and attractive.

Karen Schriver

Document design is a field concerned with creating texts (broadly defined) that integrate words and pictures in ways that help people to achieve their specific goals for using texts at home, school, or work.

Saul Carliner (chap. 2, this volume)

[I]nformation design may be better defined as:

Preparing communication products so that they achieve the performance objectives established for them. This involves:

1. Analyzing communication problems.
2. Establishing performance objectives that, when achieved, address these objectives.
3. Developing a blueprint for a communication effort to address those objectives.
4. Developing the components of the planned communication effort solution.
5. Evaluating the ultimate effectiveness of the effort.

A major reason for the varying definitions comes from the immaturity of the information design discipline and the bias of each person that reflects the previous experiences. Right now, information design can handle and should have a wide range of definitions that help spur debate and inquiry into exactly what the field does and how it should focus itself. As information design matures into its own field, an overall agreed-upon definition will emerge; one that will probably take elements from each of the foregoing definitions but will integrate them in a unique way.

The journals, research agenda, and expected practitioner actions all contribute to a discipline definition and expectations. The field has one major English language journal, *Information Design Journal*, but much of the research appears in the journals of the underlying disciplines and reflects each discipline's biases. There are several universities who have created a program in information design and I expect many more programs to appear over the next few years. All these forces, acting together over the next several years, will yield a much stronger and concise definition of information design.

Positioning information design

This section tries to place information design within the various disciplines from which it draws, as an attempt to show exactly how it fits into the "big picture." Although information design lacks an agreed-upon definition, there is less debate on which underlying disciplines feed into it. Information design "draws on many research disciplines and many fields of practice, including anthropology and ethnography, architecture, graphic design, human factors and cognitive psychology, instructional design and instructional technology, linguistics, organizational psychology, rhetoric, typography, and usability" (Redish, 1999). Of course, the proportions which each underlying discipline contributes varies widely and tends to be apparent in any individual definition.

My attempt, shown in Figure I.1, portrays information design as drawing primarily from the fields of technical communication, visual (graphic) language, and human factors. To restate, information design is not synonymous with any one of these three areas. But rather it is currently developing into its own unique discipline which draws on all three.

Visual design
Although some people consider information design as equivalent to graphic design, this book takes a much broader view. Yes, high-quality graphic design is an important part of good information design. But information design is not synonymous with graphic design or even the field of information graphics (see e.g., *Understanding USA*, Wurman, 2000). While useful, these graphics must be carefully constructed to ensure their fit with the rest of the design (Tufte, 1983). Whereas many graphic designers are excellent at developing these graphics, far too often, they come at a project with the idea of doing unique, cutting edge artwork, regardless of the communication requirements of the project.

Information design communicates content to the reader. Information designers bring together prose, graphics, and typography and make them work in unison to achieve the desired effect. Information design

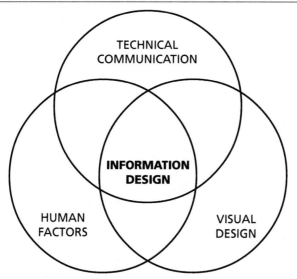

Figure I.1. Information Design as the Intersection of Different Disciplines. Within the area of overlap, a gestalt of knowledge is occurring as the information design discipline grows. A discipline that will someday be viewed as independent of its foundational disciplines.

encompasses any and all ways of clearly and effectively communicating a set of information to a reader (Horton, 1990). That might include the layout of a text-only report, the development of the user interaction for a software application, or the design of a flashy four-color report for prospective clients. The importance of the appropriate visual presentation cannot be overemphasized. It drives both the reader's emotional response and ease of manipulating the information (Carliner, 2000).

Human factors
Here I consider human factors and usability as roughly equivalent terms. The need to consider the user through the various user-centered design methods and to test that the intended responses actually happen are vital to quality information design.

The user experience is ultimately what will determine the success or failure of any information design project. User experience includes all points of interaction a user has with the information. User experience with printed documents is defined by factors such as, the paper and print quality, volume of information (2-inch thick report vs. a 4-page summary), and graphics. User experience on the web is defined by factors such as, navigation, structure, content, layout, graphics, flow, and linking strategies (Nielsen, 2000).

Ensuring that the design fits the specifications and provides the proper user experience is what human factors does. Without an adequate plan for ensuring usability, there is no guarantee that the design is anything but a perfection of communication within the designer's mind.

Effective human factors is much more than testing various designs to find which one gives the best time to find a result. In creating the optimal user experience, the information designer must also consider the social context of the user (Odell & Goswami, 1985). Often, creating an effective user experience has more to do with defining who your audience is, what they're there for, and what social, political, cultural and other conventions they are familiar with, than it is with providing a technological method of presenting information (Hackos & Redish, 1998).

Technical communication

In the final analysis, information design requires content. Although much of the information design process operates above the level of the text itself, in the final analysis, the text content must mesh with the design (Schriver, 1997). Any design lives and dies by the content it has to impart. One example can be seen in the criticism of the use of Flash for current web sites; the flashiness overpowers the information. Once a reader gets past the "cool factor," nothing is left.

The technical communicator's skills for transforming the information from its source to the proper level for the audience underlie communicating information. This skill set will always be needed to support the work of the information designer. As part of this skill set, technical communication brings the methods of aiding information communication, such as headings, text and graphic integration, and page layout.

Along with writing and editing skills, technical communication also provides the methodologies needed to define the user's needs and goals, and task and audience analysis.

Difficulties of good information design

Mark Twain declared that it takes 3 weeks to prepare a good impromptu speech. Good information design must be like a good impromptu speech. Like Twain's speech, much hard work goes into the effortless flow of information from source to reader. The design must make everything clear and functional without distracting from the information being conveyed. The hard part for the information designer is making the design disappear. Rather than being something the reader focuses on, the design must carry the information to the reader in a clear manner while remaining out of sight. As the reader's attention to the design itself increases, the amount of content conveyed to the reader decreases.

The interesting thing about good design is that most people don't realize how hard it is. Or, worse, they equate it with slapping on some color and a few graphics. Every information designer has had at least one client say, "This project doesn't really require much design. Just make it clear!" Unfortunately, projects described like this usually do require a lot of design work; a design, which when successful, is never noticed. And when it fails, the entire project fails. In a good design, readers can effortlessly extract the information they need without being conscious of how they gain the information. Consider your past experiences finding information on large web sites. Some sites have easy to understand text and structure which quickly lead you to the information you want, whereas other sites submerge you in a maze of confusing links leading from one chunk of minimal information to another. Interestingly, both sites might have comparable color schemes and pleasing graphics. The difference exists in how the site portrayed its content.

Although much of the information design literature focuses on graphic design or human–computer interaction issues, information design should never be considered the practice of web navigation, creating graphics, picking fonts, laying out a page, or using particular tools. Rather, it must be considered the practice of enabling a reader to obtain knowledge. Many different elements come into play for an information designer, but knowing about each of them does not constitute being an information designer any more than knowing how to use hand and power tools constitutes knowing how to build a house. The tool knowledge may exist, but does the person possess the contextual knowledge of how to properly apply them to the situation?

Rather than tool knowledge, the essence of being a good information designer is one of understanding the following:

- which questions to ask a client, subject matter expert, and user
- how to listen to the answers
- how to differentiate between the client's wants and needs
- how to understand the information needs within the situational context
- how to translate the needs into results

At times, buried beneath the details of a design project, it can be hard for the designer to remember that someone is going to be looking at the information product and using it for a specific purpose. People have a goal in mind when they use information; the information designer must ensure they can reach that goal with minimal problems.

When working on a design, information designers must avoid their own affinities, prejudices, and jargon, while developing a design which works given the client's needs, beliefs, desires, and behaviors. It is very easy to fall into a trap of creating a design that makes perfect sense to an "insider" but makes no sense to anyone else. Situation comedies thrive on plot twists caused by characters misinterpreting (overhearing) language that made perfect sense in the context of the original conversation. In the same way, an information designer must ensure the design does not introduce any confusing twists.

Multiple times in the previous paragraphs I've used the word information. Methods of presenting information to a reader is what this book is about. Reread the initial paragraph of this chapter and notice the focus is on *communicating information*. The media is not the concern. As I write, web sites using Flash and Shockwave are appearing all over the World Wide Web and getting slammed in the various usability and information design mailing lists. Unfortunately, these sites often forget (or don't know) that they have a message to deliver and instead are little more than a multimedia production of special effects that leaves the reader dazed and fails to deliver a coherent message. Dazed readers suffering from information overload should never be the goal of a design. Using fancy technology can be highly seductive and fun for the designer, but not necessarily the best for the reader. Communicating information to the reader must always remain the focus. In the poorly designed web sites, Flash or otherwise, I often wonder if the designers ever attempted to understand why and how people would be using the site or if the designers just took a pile of content stuff and created something that contained all of it.

Good information designers do not start with a six screen web-based design and then figure out what information to fit into it and who the audience will be. Nor do they start at the top of the menu structure and worry about how to create a design to communicate something about File-New, then File-Open, and so on. Rather, they start with understanding the information needs of the audience and what data is available, then decide if paper, web-based, or a loudspeaker is the best method of communicating that information. The medium used to communicate the message should not be chosen until the information needs of the audience are defined.

In connecting information design to context, I'll go so far as to paraphrase Norman's claim that "to the user, the interface is the system" and rephrase it as "to the user, the information content is the system." Unless that information is properly designed, displayed, and can be manipulated for interpretation, the information (and consequently, the system) is a failure, period. Hopefully, this book will provide a vital contribution to helping to design systems that contain information that is properly designed and displayed.

Overview of the chapters

Information Design in Motion—Beth Mazur
Whereas the other chapters look at the present or future states of information design, Mazur looks at its history, one which extends much farther back than the past 20 years. The chapter examines information design from both a historical and speculative perspective, describing a range of potential information design products, not just information-rich graphics (of which Minard's graphic of Napoleon's march to Moscow is perhaps the most famous example) and looks at some of the conversations that have been occupying today's information designers.

Physical, Cognitive, and Affective: A Three-Part Framework for Information Design—Saul Carliner
This chapter, the only reprint of the volume, is an article printed in *Technical Communication* (Winter 2000) which should become one of the classic articles in information design.

Carliner first explores limitations with the prevailing concept of document design and then offers a definition of information design. But more than simply another definition, the article develops that definition into a framework meant to broaden the perspective of information. That framework describes the three types of design activities involved in technical communication: physical design, cognitive design, and affective. Finally, he considers the strengths and limitations of this framework.

Collaborative Processes and Politics in Complex Information Design—David Sless
David Sless draws on his experience at the Communication Research Institute of Australia (CRIA) to provide a description of both the design methods and some short case studies. He focuses on what he has long called collaborative design methods or, the more recent term, user-focused design methods. After a discussion on the stages of the collaborative information design used at the CRIA, he discusses how following these stages helps to ensure that the real users are represented in the design. The case studies focus, not on the design process itself, but on examining the problems and political factors which have influenced various projects and how these factors can seriously impede the development of an effective design.

The Five Dimensions of Usability—Whitney Quesenbery
In common parlance, usability is often equated with ease of use, a satisfyingly simple reduction to focus on the user's interaction with the

product. This focus concentrates on user actions toward a goal. In this context, it is understandable that the role of information design in enhancing usability has been obscured. Quesenbery identifies a five-dimensional framework for connecting the usability to the user experience, which taken together, can both describe an experience and serve as a guide for design. By understanding the users, their goals, and context of use, the relative importance of each dimension can be determined. Part of the chapter explores how these five dimensions come into play in this informational world, where the content must carry much of the burden of creating a usable experience.

Applying Learning Theory to the Design of Web-Based Instruction— Susan Feinberg, Margaret Murphy, *and* John Duda

Feinberg, Murphy, and Duda report on a study in the design of a web-based instructional module as part of an interdisciplinary team project. The team's objective was to apply cognitive load theory to the design of web-based instruction and user test the product. This chapter describes cognitive load theory as it applies to the design of effective instruction. It also presents guidelines for the effective uses of multimedia and graphical user interfaces, especially as they inhibit learning and impose unnecessary cognitive demand on the learner.

What Makes Up a Procedure—Hans van der Meij, Peter Blijleven, *and* Leanne Jansen

The key part of any manual is, of course, the presence of information that supports the user's actions. Over and over again, research indicates that users are predominantly—but not exclusively—interested in this type of information, as opposed to declarative or background information. Going beyond the numerous style sheets and extensive discussions on how to present procedures, the authors undertake a systematic study to analyze and describe procedures as they appear in (a broad sample of more than 100 manuals) technical documentation. They then connect the analysis with both the theory and practice of instruction writing.

Visual Design Methods in Interactive Applications—Jean Vanderdonckt

Vanderdonckt illustrates how visual design techniques can serve for laying out information items and interactive objects in user interfaces of interactive multimedia applications. These objects are generally known and designed for their great user feedback and interaction through simple interaction objects (e.g., list boxes, radio buttons, push buttons) and interactive objects (e.g., text, image, graphic animation, picture, video motion). Thirty pairs of visual techniques are introduced by defining their opposites on a continuum ranging from harmony to contrast.

Contextual Inquiry as a Method of Information Design—Karl Smart
The purpose of this chapter is twofold: (1) to demonstrate an information
design method (contextual inquiry) as it applies to documentation
design through a case study and (2) to report on the results of the contextual design case study, outlining insights learned about users and
showing how contextual data can inform documentation design decisions. The chapter begins with a discussion of the contextual design
methodology, outlining the development team's organization and how
they determined their research focus. Smart shows how the team gathered and interpreted user data and describes the process of creating an
affinity diagram and consolidated work models.

*Dynamic Usability: Designing Usefulness Into Systems for Complex
Tasks*—Barbara Mirel
Systems that support users' complex tasks and problem-solving have
unique demands in terms of presenting users with the right information
in the right design at the right time. These systems for complex tasks and
problems must be adaptable.
   This chapter proposes a conceptual framework for conducting the
usability and user experience inquiries that are needed for first defining
and then building usefulness into systems for complex tasks from the
start. It then identifies the applied effects of these dynamic usability
inquiries, particularly stressing the ways in which findings need to shape
decisions about system architecture and scope. Next the chapter analyzes
how building usefulness into architecture and scope ultimately impacts
the information design that users see in interactive interfaces and help
systems. It concludes by addressing the political obstacles that commonly challenge usability and information designers in these efforts to
attain dynamic usability.

Complex Problem Solving and Context Analysis—Michael J. Albers
This chapter connects complex problem-solving research with situation
awareness research to define a method of developing web-based knowledge management designs which assist the user in solving complex
problems. In effective design for complex problem solving, the focus
must be on providing the appropriate content for the user's real-world
goals and information needs. Thus, the process of supporting complex
problems is to help the users (1) identify the important elements of the
situation, (2) identify the relationships between the elements, and (3)
identify the information required to ensure the decision is resulting in
the desired response.
   Content analysis, developed in this chapter, provides a framework
for ensuring that the information within a system can answer the above

criteria. It also ensures that the designer has enough situational knowledge to present them in a manner that fits Marchionini's three dimensions of information: specific to the situation, in the proper quantity, and presented in a timely manner.

Applying Survey Research Methods to Gather Customer Data and to Obtain User Feedback—Beverly Zimmerman and Maribeth C. Clarke

Software documentation writers frequently have to gather information about their customers or obtain feedback about their documentation. Much of this information gathering and usability testing is based on a written or verbal question-and-answer process that results in answers that are used to measure the quality of the software's documentation. It is important, therefore, for documentation writers to understand how to create reliable and valid measures. This chapter reviews recent work in survey research and summarizes the principles documentation writers should know to gather usable data about software users and to create effective measures of their documentation.

Single Sourcing and Information Design—Ann Rockley

People often have to create documents for different audiences and for different media (e.g. web, print). However, because timelines and budgets for developing information are often tight, we need more efficient ways to develop information. Single sourcing is a method that can address all these needs. Single sourcing enables you to create information for multiple users with multiple needs and build customized documents "on-the-fly." Although single sourcing does take more up-front planning, it can significantly decrease costs and development times once implemented. This chapter describes single sourcing, its benefits and costs, and provides a clear process for developing effective single source materials.

Redesigning to Make Better Use of Screen Real Estate—Geoff Hart

Developers often ask writers to help them fit all the necessary text into a dialog box or other component of the user interface. One common request is to reduce the labels of the interface elements to no more than "two or three words." This chapter proposes an iterative strategy for analyzing the problem, and presents two case studies that demonstrate application of the principles. Careful reexamination of relationships between elements of the information and redundancy in how those elements are presented, combined with knowledge of the sequence users will follow to actually use the information, often reveals simple solutions for resolving the problem of limited space.

References

Carliner, S. (2000, May). *Document, information, and communication design.* Society for Technical Communication Annual Conference. Orlando, FL.

Hackos, J., & Redish, J. (1998). *User and task analysis for interface design.* New York: Wiley.

Horton, W. (1990). *Designing and writing on-line documentation.* New York: Wiley.

International Institute of Information Design. (2000, November 26). members.magnet.at/simlinger-iiid/English-2.html

Information Design Journal. (2000, November 26). www.benjamins.nl/jbp/journals/Idj_info.html

Jacobson, R. (Ed.). (1999). *Information design.* Cambridge, MA: MIT Press.

Norman, D. (1986). Cognitive engineering. *User centered system design: new perspectives on human-computer interaction* (pp. 33-62). D. Norman & S. Draper. (Eds.). Hillsdale, NJ: Lawrence Erlbaum Associates.

Nielsen, J. (2000, December 4). Alertbox. www.useit.com/alertbox

Odell, L., & Goswami, D. (1985). *Writing in nonacademic settings.* New York: Guilford Press.

Redish, J. (1999). What is information design. *Technical Communication* 47(2), 163-166.

Schriver, K. (1997). *Dynamics of document design.* New York: Wiley.

Taylor, C. (1999). Information design: A European perspective. *Technical Communication 47(2),* 167-168.

Tufte, E. (1983). *The visual display of quantitative information.* Cheshire, CT: Graphics Press.

Wurman, R. S. (1989). *Information Anxiety.* New York: Doubleday.

1

Information Design in Motion

Beth Mazur
University of Baltimore

In the foreword of *Visual information for everyday use*, Paul Stiff (1999) looks for "the *proper* history" which "is written of the thinking and practices which cluster around information design—an awkward term for a still contested idea" (p. xi, italics added).

This chapter is not that history. A "proper" history is the subject of its own book, perhaps requiring, as Karen Schriver (1997) suggests, a historian to do the subject justice. As such, this history is just one of "many histories of information design that could be constructed" (Sless, 1998, p. 3). My apologies, in advance, for omitting any important facts or people.

Ultimately this chapter is part bibliographic essay, part speculation. It provides a historical context, a look at where information design is today, and where it may be going—information design in motion.

A brief history of information design

In this section, I touch on early work in the field, look at the field's formative years (1970s and 1980s), and then discuss how the Internet has played an important role in information design.

Early work in information design

One of the challenges in developing a history of information design is first choosing what one pays attention to. As the introduction to this volume suggests, what information design is depends considerably on how one has come to the field. Given this, a history of information design naturally depends on one's perspective (or perspectives)—or on one's familiarity with different kinds of information design products.

Information graphics
As perhaps the most well-known product in information design, consider the work of Edward Tufte and Richard Saul Wurman. Tufte and Wurman have gained a lion's share of the press for their work in the field (at least in the United States), resulting in a natural tendency to view their work as representative of the entire field of information design.

Tufte's perspective is that of a statistician, with an emphasis on data-rich information graphics. A famous example of this approach would be Charles Minard's 1861 diagram of the losses incurred by Napoleon's army in the Russian campaign of 1812, of which Tufte (1983) notes: "it may well be the best statistical graphic ever drawn." (p. 40)

Tufte (1983) credits William Playfair (1759-1823) with developing or improving upon "nearly all the fundamental graphical designs" (p. 9) In later books, he draws from even earlier "information designers," commenting that the "wonder of Information Design is that I can write a book in 1990 and the main intellectual hero is Galileo" (Computer Literacy Bookshops, 1997). In this same interview, he credits John W. Tukey, "the phenomenal Princeton statistician," with opening his eyes to the importance of the field of statistical graphics in the mid 1960s.

Robert Horn, the founder of Information Mapping,[1] also approaches information design from primarily an information graphic perspective. He described the different approaches to information design as information graphics, presentation or business graphics, scientific visualization, interface design, or wayfinding. (Horn, 1999) Thus it is no surprise that for him, the roots of the field are similar to Tufte's. He credits Playfair, and acknowledges Florence Nightingale for "inventing new types of statistical graphs and being one of the first to use information design in a public policy report" (p. 17).

He extends this history in *Visual Language* (Horn, 1998) by providing a timeline of the "innovations" in visual language, starting first with Egyptian hieroglyphics and concluding with the World Wide Web. In addition to Playfair and Nightingale, he credits early pioneers such as Joseph Priestley as the inventor of the biographical timeline (1765), Michael George Mulhall as the inventor of pictorial statistics (1885), Otto Neurath as the inventor of ISOTYPE, the use of pictographic visuals for information graphics (1930s),[2], and Henry Gantt as the inventor of the now well-known Gantt chart for project management (1900–1911). [3]

Obviously all of these are important precursors to the field of information design today. Yet they are only part of the picture, simply because one can do information design without graphics—and certainly without the rich information graphics described by Tufte and Wurman.

Public information products
Another type of important information design product are the "public information products" (Easterby & Zwaga, 1984) such as instructional text, tax and insurance forms, and medical leaflets. Information designers working on this type of product typically focus on creating usable products through the use of graphic design elements such as typography and layout. An excellent source for early work in this field can be found in the journal *Visible Language*[4] (published as *The Journal of Typographical Research* from 1967 to 1972).

Karen Schriver (1997) made note of the following early design pioneers: Edward Johnston as the creator of the sans serif typeface for the London Underground (1916), Ernst Keller as an early proponent of the grid system of design (1918), Walter Gropius as the founder of the Bauhaus movement (in 1919) with its emphasis on "the functional use of grids, asymmetrical organization of elements, and sans serif typefaces" (p. 110), and Jan Tschichold as the author of *Die Neue Typographie* (1928).

Documents
This third set of information products are closely related to the second; they differ primarily by an increased emphasis on writing and rhetoric (Schriver, 1997).[5] This group of products typically include those familiar to technical communicators: user manuals, reference guides, or online help. Early pioneers in this area include: T. A. Rickard, who published *Guide to Technical Writing* (1908), Samuel Chandler Earle, who published *The Theory and Practice of Technical Writing* (1911), becoming known as "the Father of Technical Writing Instruction" (p. 108), and Kenneth Burke, who published *Rhetoric of Motives* (1950).

Wayfinding, maps
An important variant of public information products are those that are meant to assist readers with finding their way in physical space— "wayfinding" or "environmental communication." (Passini, 1999, p. 88) Wayfinding products include road signs, traffic-management pictograms, airport and railway station signage. These types of products have been particularly important in Europe, with its large, nationalized transportation organizations and multilingual populations.

In *Visual Function*, Mijksenaar (1997) described how early designers, influenced by the Bauhaus' abolition of capital letters, chose to use all lower-case letters when Amsterdam's Schiphol Airport was constructed—"despite the fact that legibility research conducted in 1960 had shown what every typographer had long known, that the recognizability of names, especially in the kind of search operations involved in reading signposts and forms, increases significantly when each name or sentence begins with a capital letter" (p. 22).[6]

One of the most striking examples of early information design in wayfinding is the work of Jock Kinneir and Margaret Calvert in the 1950s and 1960s for the British road network, where experimental work in perception and cognition were critical components of the project (Kinneir, 1984).

The formative years

The previous section described some of the products of early pioneers of information design. My categorizations of these efforts are likely to be debated, but one thing is certain; none of these early pioneers would have described their work using the label *information design*. Information design as a distinct field began to emerge in the late 1970s, although the term appeared a bit earlier. In *Graphic Design for the Computer Age*, Edward A. Hamilton (1970) wrote prophetically that the "information designer is in reality a teacher—and no better compliment may be paid a professional designer than to call him a teacher" (p. 14).

The more widely recognized origin of modern information design was the 1978 NATO Conference on Visual Presentation of Information held in Het Vennenbos, The Netherlands. The conference took an interdisciplinary approach, involving cognitive psychologists, human factors researchers, engineers, typographers, and designers to "try to relate visual and perceptual research to the practical problems of designing information displays" (Easterby & Zwaga, 1984, p. xxii).[7]

Rather than deal with the phenomenon of "heavily oversubscribed" (p. xxi) sessions related to computer applications, the Het Vennenbos conference instead focused on simple, public information products such as signage, forms, and procedural aids. Another important objective was an emphasis on the interdisciplinary aspects of information design.

Easterby and Zwaga subsequently undertook the effort to publish the contributions of the conference participants. The resulting work was published in 1984 as *Information Design: The Design and Evaluation of Signs and Printed Material*, with contributions organized into sections on theory and method, design parameters, and multiple sections on applications in sign systems, road traffic signs, consumer/safety signs, and printed material (such as procedural manuals and forms).

Although the conference is a notable event in the early years of information design, perhaps the major legacy of the conference was *Information Design Journal* (IDJ). As Easterby and Zwaga (1984) note, it was at Het Vennenbos that "Rob Waller gained support ... for his intended *Information Design Journal*, which ... numbers among its editorial board many of the experts who were at its conference" (1984, p. xxv).

In fact, the IDJ actually predated the publication of the 1984 Het Vennenbos volume; the first issue of IDJ was published in 1979. Rob Waller (1996) wrote:

> IDJ was started to consolidate a community of interest—an invisible college—that had emerged in the 70s among a number of designers, teachers and researchers. It built on and was inspired by Michael Twyman's curriculum at Reading University Typography Department, Merald Wrolstad's *Journal of Typographical Research* (later renamed *Visible Language*) and the interdisciplinary work of Herbert Spencer and his RCA colleagues, Patricia Wright at the MRC,[8] Jim Hartley and Peter Burnhill, and the Textual Communication Research Group at the Open University, where I was based. IDJ had a definite agenda—to get specialists in language and design talking to each other, and to make research more accessible to designers.

Waller (1996) noted that he and his coeditor, Bryan Smith, "tossed a number of terms about before settling on [information design], but whether it was original or not I can't say."[9]

The inaugural issue of the IDJ included articles on teletext and viewdata (Linda Reynolds, Royal College of Art), design history and the visual language of design (Bryan Smith, London College of Printing), quality control of document design (Patricia Wright, Medical Research Council), and functional information design (Rob Waller, The Open University), among others (*Information Design Journal*, 1979). Over its early years, the IDJ also organized five conferences for its readership, thus providing an opportunity for members of this community to meet and share ideas (Waller, 1996).

As a journal with an international readership, the IDJ deserves much of the credit for making information design visible as a field. But its efforts were greatly assisted by the launch of the London-based Information Design Association[10] (IDA) in 1991. For the first half of the 1990s, the IDA organized meetings and generated a newsletter, *IDeAs*,[11] for its membership (Waller, 1996). Although the IDA may have not achieved some of the lofty goals of its founders (see Waller), the IDA clearly provided considerable momentum for the development of the field.

Up until this point, I have been describing activities centered primarily in the UK and the Netherlands. But there were also others involved in early information design work. In fact, another information design group, the International Institute for Information Design (IIID),[12] was formed in 1988 and centered in Austria, although it was fairly inactive until the mid 1990s.

Others included the Communication Research Institute of Australia, whose early activities are described by David Sless (1998), the design work of Eric Spiekermann in Berlin (president of the IIID, founder of MetaDesign, and coauthor of the popular typographic volume *Stop Stealing Sheep & find out how type works*), and that of Rune Pettersson of Mälardalen University in Sweden, author of *Visuals for information, research, and practice* in 1989 and vice president for the IIID.

On the ID-Café, Conrad Taylor (2000) described the work of Sven Lidman, the Swedish encyclopedia editor:

> He's been a strong advocate of what Bob Horn calls "Visual Language," but for which Sven coined the term "lexivisuals" some twenty or more years ago. He founded the Bild och Ord Akademin (Picture & Word Academy) and financed a kind of Nobel Prize for information design in Sweden (Lidmanpriset), and has been a strong advocate of replacing text with explanatory graphics in textbooks, newspapers, magazines etc.

Meanwhile, across the pond in the United States, there were complementary activities, even if they weren't described using the term *information design*. Schriver (1997) described in detail the work of the Document Design Project (DDP), an effort funded by the National Institute of Education. The project involved academia and industry, including the American Institutes for Research (and its Document Design Center, led by Janice Redish), what became the Communications Design Center (CDC) at Carnegie Mellon University, and the design firm Siegel & Gale.

Schriver (1997) noted: "The output of the DDP was rather impressive. The Project provided training to government personnel from more than 15 federal agencies … on at least 13 different projects" (p. 73). In addition, the work on the DDP produced two foundational works, *Review of the Relevant Research* in 1980, and *Guidelines for Document Designers* in 1981.

Both the CDC and the Document Design Center (later the Information Design Center) contributed considerably to document design research (Schriver, 1997). Unfortunately, the CDC was closed in 1990 as part of a change in institutional support; the Information Design Center was similarly closed in the late 1990s. But the work accomplished by these groups, and those influenced by this early work, led to a strong American movement toward the concept of document design, which Schriver (1997) defines as the "act of bringing together prose, graphics (including illustration and photography), and typography for purposes of instruction, information, or persuasion" (p. 10). Clearly document design is very close to information design, and there was and continues to be considerable sharing between the two communities via journals, conferences, and other related activities.

Another influence during these years was the work of American designers, such as Jan White (1982), whose *Editing by Design* provided a "how-to" approach for effective word-and-picture communication for editors and designers. White was an early proponent of the idea of design as being more than "good looks." Instead, he noted that design "is an arm of editing, that is, interpreting the meaning of, a story" (p. 3). Although he worked primarily in magazine design, his techniques extended beyond that medium to other print works.

Meanwhile, Tufte published *The Visual Display of Quantitative Information* (1983), with *Envisioning Information* (1989), and *Visual Explanations* (1997), along with writing other books and doing popular seminars on information design.

Richard Saul Wurman also created a number of information design[13] pieces during this time, such as his Access Guides (starting in 1980) and his USAtlas—the inspiration for the latter being that "you do not drive across the United States alphabetically." (1996, p. 31)

Wurman followed these pieces with *Information Anxiety* (1989), *Information Architects* (1996), *Understanding* (1999), and the revised *Information Anxiety 2* (2000). While not universally acclaimed (e.g., see Sless, 2001), they provide very interesting and innovative examples of approaches to complexity in information products and are regarded as essential works for information designers (Albers & Lisberg, 2000).

The Internet years

The mid-1990s and the increased popularity of the Internet and the introduction of the World Wide Web would impact information design in ways that could not have been anticipated. For one, this technology enabled communication between disciplines and communities that had only been marginally possible prior to this point (Schriver, 1997; Saul Carliner, personal communication, 2001). In addition, the exponential growth of web sites would cause many information designers (or would-be information designers) to begin to explore how the field could help with these new products.

In September 1994, many of the original participants of the Het Vennenbos reconvened in Lunteren, The Netherlands. The conference again focused on information design in the public information space, with emphasis on the "cooperation and dialogue" between researchers and designers (Stiff, 1995, p. 65; Nijhuis & Boersema, 1994). And as with the earlier conference, there was a lasting—and extremely valuable—legacy besides the papers and subsequent book: the InfoDesign and InfoDesign-Café email discussion lists.[14] With these lists, anyone with an email account would be able to discuss the field with others interested in the subject, share news about research, papers, and related conferences, and, in essence, build a global community of information designers.

Around the same time, the IIID became more active, as noted by director Peter Simlinger:

> We ... started with two conferences in 1993 and 1994 to decide on definitions and a work program. In 1995, we organized the first Vision Plus symposium "Designing for electronic communication" which at that time was embedded in an advanced studies course, directed by Prof. David Sless of Australia and advised by Dieter Willich, Germany. Its staff included additional experts of Great Britain, The Netherlands, and Austria. Students came from Japan, Australia, Britain, Germany, and Austria. (personal communication, 2001)

In early 1996, a U.S.-based information design group was created to support information design activities. In its initial solicitation letter to potential members, Wes Ervin described the U.S.-based Information Design Association (not to be confused with the earlier formed, UK-based IDA) as an organization "dedicated to promoting the practice of information design in North America." Its advisory board included Richard Saul Wurman as a "prominent information designer" (personal communication, March 16, 1996).

The U.S.-based IDA was affiliated with the IIID; the "close to 100" charter members received copies of the IIID newsletter, *ID News*. However, despite encouraging support from its new membership, the U.S.-based IDA was unable to sustain itself, and in 1997 merged with the IIID.

In January 1997, I founded the Information Design special interest group (ID SIG)[15] within the Society of Technical Communication (STC). With STC's membership base,[16] the ID SIG quickly grew to its current membership of about 3,000. Its activities over the last few years have centered around a regular newsletter, *Design Matters*, SIG participation at STC's annual conference in May, and other activities (such as this volume).

Information design in the present

As of the writing of this chapter, the UK-based IDA is being reorganized.[17] The IIID is working to establish IIID.Japan and develop two "thematic networks" on manual design and knowledge presentation. The STC ID SIG continues publishing its newsletter, coordinating information design activities at its annual conference, and working closely with members of the international information design community.

The *Information Design Journal* (IDJ) has been reorganized under the editorial leadership of Piet Westendorp and Karel van der Waarde (Delft University of Technology, The Netherlands) and published by John Benjamins in Amsterdam.[18] The content is now organized around a section of articles that is theme-based and a section of articles related to a recent information design conference. In the first issue of this new collaboration, the theme focuses on a discussion of Jacques Bertin's theory of geographic visualization; the conference review was of the IIID's Vision Plus 6, held in Vienna in July, 1999 (Westendorp & van der Waarde, 2000/2001).

The InfoDesign and InfoDesign-Café mailing lists underwent their own organizational change. Maintained for the first 3 years by Yuri Engelhardt at the University of Amsterdam, the lists were administered for a short time by Conrad Taylor (deputy chair and newsletter editor for the IDA).

At the present time, the lists are administered and moderated by Karel van der Waarde (who is assisted by Conrad Taylor, Karen Schriver, Alan Davis, Yuri Engelhardt, and Piet Westendorp). InfoDesign continues as a moderated list with relatively low volume, but high quality content (book reviews, conference announcements, etc.). InfoDesign-Café is not moderated and thus remains "home to many freewheeling discussions" (www.informationdesign.org).

Defining information design today

Much has happened to information design since the 1970s and 1980s. Yet one thing that hasn't changed is that people inside and outside the field are still asking the question: what is information design? As described in the introduction of this volume, there are as many definitions as groups.

From the STC ID SIG's perspective, it was important that our perspective of information design wasn't redundant with the broader concept of technical communication (or there would be no point in a special interest group within a professional association of technical communicators). We were very much influenced by Wurman's *Information Architecture*, although we chose to retain the term of the European community. Our position has been that the field of information design applies traditional and evolving design principles to the process of translating complex, unorganized, or unstructured data into valuable, meaningful information.

Or should we say, "representations of information." In his concluding chapter in *Information Design*, Jef Raskin (1999) is "delighted" to point out that information design is a "misnomer. Information cannot be designed;

what can be designed are the modes of transfer and the representations of information ... it is important for designers to keep the concepts of *information* and *meaning* distinct" (emphasis in the original)" (p. 342).

However, ultimately the term information design is just shorthand—as Wurman said, a way of "making the complex clear." Or as Conrad Taylor noted: "Using the term *information design* as a functional label for what we do doesn't preclude having some pretty sophisticated views about how information is parlayed into meaning" (personal communication, 2001).

I expect that questions of defining information design will be resolved eventually, particularly as there are so many concepts in common between the existing definitions. In the remainder of this section, I look at two other discussions within the community that have interested me recently (in fact, have been the subject of two of our newsletters). One is the question of how information design relates, overlaps, or is distinct from other fields, particularly information architecture. The other is the question of whether information design is a craft, a profession,[19] or some other kind of entity.

Many within the community are not as interested in such discussions. Indeed, Wurman noted that he found such discussions "academic and pointless" (2001, p. 9). I am sympathetic to such concerns, for as Wurman also suggests, perhaps there are better things to do with our time, given the amount of bad design! Yet, I think that there are two main issues that make these discussions relevant.

First, there is the argument that if we can't adequately explain what it is we do, how do we expect to get clients to pay for it? Information design, graphic design, technical writing, and usability are all examples of fields that may be perceived by business as being *add-ons*, things to do only when the budget or schedule permits—in other words, nonessential. The majority of practitioners in these communities know that just the opposite is true—yet efforts to provide justification (through research or case studies) showing return on investment are patchy at best.

This ties in to the second argument. Often this kind of justification effort is located within academic communities. But information design is only now becoming an established discipline within academia. What is it that these programs teach? Is information design taught at Reading University the same as the field taught at Georgia Tech? Should it be? Are the best located within traditional departments (such as design or English)? Or in new, interdisciplinary departments? Adopting a laissez-faire attitude towards what we do and how we do it—even if there are so many important things to do—seems less likely to lead to the kind of strong, sustained movement needed to make the changes that so many information designers consider important.

Information architecture or information design?

So what about information architecture? According to Wurman, he coined the term in 1975: Information architects "make the complex clear; they make the information understandable to other human beings" (2000, p. 23). This sounds very much like what could be considered information design. Yet in a post on the InfoDesign mailing list, Wurman clarified his choice of terms:

> I selected the term information "architect" rather than information "designer" as the term "designer" continues to be interpreted by the public as an individual who is hired to come in after the fact to make some project "look better"—as opposed to a professional part of the initial team creatively solving a problem.

> I do not believe I can change this popular preconception. I believe the term information architect evokes a rigor in the creation, research, choice as well as the presentation of information in an understandable yet artful form. (WWW, 3 Apr 1998)

It might have been well and good to have two different names for essentially the same concepts. However, a new group has recently staked a claim to the term *information architecture*, and as popular as Wurman's usage might be, this newer information architecture has proven to be a very influential phenomenon.

This perspective is very useful given the growing complexity of web sites, which can now contain thousands, or hundreds of thousands, of individual pages. Understanding structure, organization, and the concept of meta information to facilitate searching and navigation is a fundamental requirement for such large sites.

This new field is described by books such as *Information Architecture for the World Wide Web* (Rosenfeld & Morville, 1998) and conferences such as the ASIS (American Society for Information Society) *Summit 2000: Defining Information Architecture*. According to Rosenfeld and Morville (1998), the information architect:

- Clarifies the *mission* and *vision* for the site, balancing the needs of its sponsoring organization and the needs of its audiences.
- Determines what *content* and *functionality* the site will contain.
- Specifies how users will find information in the site by defining its *organization, navigation, labeling,* and *searching systems.*
- Maps out how the site will accommodate *change* and *growth* over time. (p. 11)

The STC ID SIG published a dialog entitled "What's in a name?" where participants were asked to discuss informally any differences between information design and information architecture. Many of the responses were unsure that focusing on labels was important—indeed, Nathan Shedroff (2001b) shared Wurman's sentiment: "I find the hoopla around the terms to be not only a distraction but a waste of time" (p. 6).

However, I found one response particularly interesting, as I'm compelled by the concept of a different emphasis on presentation and structure in information design compared to (the newer) information architecture. Jesse James Garrett (2001) wrote: "information architecture and information design are indeed quite different. ... Information architecture is primarily about cognition ... Information design is primarily about perception. ... Information architecture belongs to the realm of the abstract, concerning itself more with structures in the mind than the structures on the page or screen. Information design ... couldn't be more concrete, with considerations such as color and shape fundamental to the information designer's process" (p. 3).

As of the writing of this chapter, this discussion has not been resolved (nor do I expect it to be any time soon). As a matter of fact, the IA community itself is concerned with similar activities (Dillon, 2001; Rosenfeld, 2001; Wodtke, 2001).

The exponential growth of this web-focused information architecture has been somewhat tempered by the economic failures of many Internet ventures in 2001 (the so-called "dot.bomb" phenomenon). This is likely just a temporary set-back; there is certainly optimism for the future among that community (Dillon, 2001; ACIA, 2001). In any case, information architecture will likely continue to overlap, inform, and provide context for information design for the foreseeable future.

Craft or profession?

Another ongoing discussion is the question of whether information design is a craft, a profession, or some other kind of field. This is a natural extension of the "what is information design?" discussion, but it extends it in the direction of how people actually *practice*.

In 1999, the STC ID SIG solicited comments about the prospect of information design as a profession. Among the many interesting responses was one from David Sless (1999), who thought of information design more as a craft than as a profession: "I think sociologically and culturally we have more in common with craft workers in previous centuries than we have with professionals in our own century" (p. 8).

Conrad Taylor (2000) was similarly unconvinced that information design is a profession, seeing it "more as a stance that one takes, like a political or moral stance. ... [We] are making a promise to ourselves and each other to improve the quality of communication" (p 168).

In considering the prospect of information design as a craft, Jacobson (1999) noted that information designers may apprentice, just as carpenters or masons, while suggesting that there is still room for science to avoid "shoddy workmanship" (p. 6). In reviewing Tufte's principles for designing graphics, Raskin (1999) noted that the last one, "'revise and edit,' tells us not only to check repeatedly that the first four conditions are met, but also to apply our aesthetic judgment to the final work" (p. 345). In other words, there is some art along with the science.

My interest in this question is certainly not to suggest that information design be organized as a profession like medicine or law, with rigid roles, education, or certification procedures. This discussion has been going on in the technical communication community (Davis, 2001; Hayhoe, 2000; Mead, 1998) for a long period of time without resolution. It certainly isn't my intention to suggest this path for the information design community (especially as I don't support it for technical communication, my home perspective).

Rather, I do think that understanding more what information designers do (and how they do it) is certainly a precursor for helping those who are interested in this field. How do they learn? Is book learning sufficient? (If so, then the field is clearly in need of a more useful "how-to" book!) Is a single course sufficient? Perhaps a certificate program? Or is coming out of a more traditional field such as graphic design or technical writing sufficient, as long as one "apprentices" to an information design-oriented design consultancy or similar company?[20]

What's more important: design or usability?

The connection between information design and the web-focused information architecture is an obvious one. But there are many other fields that touch information design. Rune Pettersson (1998) suggested that information design "encompasses more than fifty established disciplines and areas of research" (p. 67). It is beyond the scope of this chapter to describe each of these in detail. But there are a number of fields whose relationship with information design might benefit from a more thorough exploration: visual thinking/visual literacy (e.g., Arnheim, 1969, Dondis, 1973, Kostelnick & Roberts, 1998); information visualization (e.g., Card, Mackinlay, & Shneiderman, 1999); and interaction design (e.g., Cooper, 1995, Mok, 1996, Shneiderman, 1997, Shedroff, 1999).[21]

Perhaps one of the most important of these related fields, not yet explicitly listed above, is usability. Like information design or information architecture, usability is an interdisciplinary field. Its practitioners are likewise involved in discussions of what they do, what is effective, and their role in the overall design and development process (rather than as an add-on activity at the end of the process).

Most information designers consider usability to be an important component of their work (see Adams, 1994 for a specific discussion of usability and information design). Yet historically the two communities have evolved separately, with different professional association, journals, and discussion lists. And recently, they have been portrayed at being in conflict (Cloniger, n.d., Cohen, n.d.).

Part of this "conflict" has been fueled by seemingly rigid guidelines by usability gurus such as Jakob Nielsen, whose name dominates popular reporting on the subject. For example, Nielsen (n.d.) continues to recommend that web pages "weigh" less than 30kb, which "rules out most graphics." Guidelines such as these are enough to strike fear (or disdain) in many designers. On the other hand, the vision of the graphic designer as being motivated more by visions of industry awards lining their walls provides the basis for a similar set of emotions among usability specialists.

Alas, this latter vision is fueled by some gurus in the field of design. For example, MetaDesign's white type on "retina-frying" red is certainly curious from a usability perspective (www.metadesign.com). Similarly, I'm not exactly sure what 'experience' the experience designers at the Advance for Design (advance.aiga.org) had in mind with their striped home page design. And Wurman's latest book, *Understanding USA* (2000), is likewise curious…a book without page numbers, table of contents, or an index, that offers up examples such as low-contrast type on top of a photo of shifting sands (by designer Ramano Rao).[22]

Like many information designers, I prefer to focus on how usability research can inform good design, just as those involved in the field in the formative years have done. Perhaps information designers would be well served by getting more examples of usable design into the popular press. It certainly isn't that this type of work doesn't exist. For example, 20 years ago, the American Institutes for Research's *Guidelines of Document Designers* (Felker, Pickering, Charrow, Holland, & Redish, 1981) described usability-focused guidelines in four categories (organizing text, writing sentences, typography, and information graphics). Or for a recent example, there is *Designing Visual Interfaces* (Mullet & Sano, 1995).

Thus the answer to the title of this section: "What is more important, design or usability?" is neither—they are both important and the best products integrate usability *and* design in a more collaborative, holistic way. Perhaps information designers, with their interest in both aspects, are well suited to bridging the gap between traditional designers and their usability counterparts.

The future: New products, new perspectives

Given where we have been and where we appear to be currently, what might be the future of information design? For one, Karen Schriver suggests that we're poised to rescue people from the "wretched information organizations put out" as "consumers realize they 'don't have to take it anymore'" (Hunter, 2000/2001, p. 29). A *Network*-esque[23] vision of people opening their windows and tossing their poorly written software manuals and tax forms out the window may be farfetched, but surely the momentum is gaining for a world view that no longer sees design as cosmetic, as something to do only when the budget or schedule allows.

Creating better products is certainly a worthwhile task, and one that may occupy information designers for the foreseeable future. However, there may be other opportunities of interest to information designers.

Information design and the Web

One thing seems true of the "modern" information architects—they are primarily interested in solving problems on the World Wide Web. However, this doesn't necessarily mean that one can't be an information designer and do similar work.

Christina Wodtke (2001) described a model of information architecture that consists of three parts: content architecture, interaction design, and information design. The latter "concentrates on both organizing information for comprehension but also concerns itself with [GUI] design," with content architecture describing more structural aspects (described best by Morville & Rosenfeld, 1998).

In a graphic entitled "IA Areas of Practice," the authors describe three areas of practice: users, context, and content. (Rosenfeld & McMullin, 2001) It's not clear where or how "traditional" information design might fit into such a model. But as suggested earlier in this chapter, people tend to derive solutions to problems based on their background.[24] Information designers interested in web design need to communicate more with their information architecture peers and provide their own models for usable design.

Certainly the Web is a tough challenge for those who are used to the structured world of print. Designers who choose a particular typeface, line length, or color scheme can find the Web a daunting challenge, especially given the ability of the user to change many items at will.

Yet it remains that issues of usability, legibility, and readability (among others) cannot just be glossed on to an otherwise strong foundation. For reasons of cost and efficiency, more governments, non-profits, and corporations will be making information and services available on the Web.

Assisting people with navigation tasks through multitudes of pages is certainly a worthwhile task; so is helping people make use of the information once they find it. Whether this is the work of one, two, or many fields or disciplines remains to be seen. But information designers surely need to be part of the discussion.

Information design and multimedia

As Westendrop and van der Waarde (2000/2001) suggested in their editorial, information design may well begin to concern itself more with "sound, animations, and real-time video" (p. 1). Indeed some have already begun to explore this world (Barass, 1996; Thwaites, 1999).[25]

Although these multimedia products will never replace traditional print products (or traditional audio and video products for that matter), the possibilities for this use on the Internet and/or WWW can be compelling. Asked about the prospects of broadband, Sun Microsystems CEO Scott McNealy suggested that distance learning and telemedicine are two likely applications for this technology.[26]

The key, however, is that multimedia products live up to their promise, unlike that of multimedia CD-ROMs of the 1980s and early 1990s (Shedroff, 2001a) or the virtual reality of the early 1990s (Laurel, 1993).

Experience and engagement in information design

Since 1998, the field of *experience design* has been taking form (centered primarily within the graphic design community; see advance.aiga.org). As a field in its relative infancy,[27] definitions of experience design can make grasping the concept somewhat difficult: "Experience design is the way in which meaning is communicated in the network society, where no point of contact has a simple beginning and end, and all points of contact must have meaning embedded within them" (Advance for Design, 2001). In *Experience Design 1*, Nathan Shedroff (2001a) noted that it is "so new that its very definition is in flux."

Information designers certainly can't fault their experience design peers for lack of a single, strong definition. Yet because of this "flux," it remains to be seen how experience design may influence information design. As Karen Schriver (2001) noted:

> Experience design ... takes us beyond traditional user-centered design (UCD). In the traditional model of UCD, the main concern was helping people understand. People were considered in terms of their thinking and performance. In contrast, newer models such as experience design recognize the need to consider people's thoughts and feelings, that is, the interaction between

cognition and affect. But the activities that comprise "designing experiences for people" do not need the label experience design to define them. (p. 8)

One of the challenges in envisioning how experience design impacts or extends information design can be found analyzing the current products of experience design. Shedroff's book, *Experience Design 1*, will likely elicit many of the same critical comments from the information design community that *Understanding USA* did.[28]

It is certainly true that books such as this one (from an academic press) are short on design; there is only so much Times Roman that one can take. However, the pendulum seems to have swung the other way; some books seem directed more at a weaned-on MTV audience that requires constantly changing visual stimulation in order not to be bored.[29] Designing for multiple audience needs remains a most challenging task.

Certainly some information designers have recognized the role of experience. Erik Spiekermann (1995) noted: "Good information design must communicate by convincing us, not just browbeating us. And information designers would do well to keep something else in mind. They need to know, as clever advertising people have long known, that nothing convinces people more than being entertained."

In other words, in an otherwise "good" information product, per-haps there is room for design that is engaging, [30] entertaining, or inter-esting—if for no other reason than as a market differentiator. Of course, the caveat to this is that this engaging or entertaining in the reader's eye, as opposed to being only in the designer's eye.

The future: New roles

Another activity may be to explore new roles for information designers. Consider the implications given in the idea of *broad* and *narrow* informa-tion design. Redish (2000) defined broad ID as "the overall process of developing a successful document" and narrow ID as "the way the information is presented on the page or screen" (p. 163).

A similar distinction was made by Peter Morville (2000) for the field of information architecture:

At one end of the spectrum, the *little information architect* may focus solely on bottom-up tasks such as the definition of metadata fields and controlled vocabularies.

At the other end, the *Big Information Architect* may play the role of "an orchestra conductor or film director, conceiving a vision and moving the team forward," as described by Gayle Curtis.

The information designer as director

This idea of a broad role particularly resonates with me, as I am intrigued with the idea of the "information designer as director."[31] As information products become increasingly complex, involving text, graphics, sound, and/or motion video, will information design—or some variant of it—become more and more like making a film?

Dale Spender (1995) noted that the "days when the author could 'write' simply in the medium of print are numbered. ... Multimedia will be taken for granted by the electronic initiates. They will want words and sounds and images. Their narratives will be less like a book, more like a film" (1995, p. 81). Similarly, Brenda Laurel (1998) noted that it may be likely that the designer of interactive systems "will be a team of individuals who, like the playwright, director, actors, technicians, and scenery, light, and costume designers in the theatre will contribute different skills toward the realization of a common vision" (p. xiii).

Naturally there will still be a need for "traditional" information design products. But what happens when the complexity of the product increases? In this world, perhaps the "little" information designer is the specialist; the one who may be concerned with issues of typography, sentence length, or type of information graphics. It might then be the "big" information designer who ensures that the vision for a large or otherwise complex product is carried out over the lifetime of the project. He or she might be the person who facilitates communication between specialists who "speak" different languages of design, of technology, of content. He or she is the director.

Others have described this concept of oversight or management in the world of information design. Zwaga, Boersema, and Hoonhout (1999) noted that the "aim should be the integration of all design factors ... This requires a management approach to design. A designer is a product development manager who coordinates the contribution of all the different experts and is the custodian of the budget and the time schedule" (p. xxxi).[32] Indeed, back in 1970, Hamilton (1970) said, "A *designer* is not necessarily the man at the drawing board wielding pen, brush, and ruler; most often, he is the *planner* and *initiator*" (p. 14). This is information design as an overarching process.

Designing information design

In this chapter, I've provided some historical context, along with some of the discussions that have engaged people in and out of the field. But like the Indian fable of the six blind men and the elephant,[33] I may well be missing other important and interesting perspectives. However, the following is valid today as when I wrote it originally:

> The future of information design looks very promising. As Paul Sagan of *Time* said, "No one wants to sit at the bottom of Niagara Falls with a bucket, saying 'I can't keep up with all this.'" A future as an information designer is not just a good career choice; it can really make a difference in people's lives. (Mazur, 1999, p. 2)

Information design is in motion. It is being shaped by the activities of those participating in the field. Information design is likely to change based on new ideas coming from research and practice. Although there may be disagreements or conflicts over what ideas will gain momentum, ultimately I expect that information design will emerge as a stronger— and far more visible—discipline.

Acknowledgments

Many thanks to Karen Schriver, Peter Simlinger, Conrad Taylor, and Karel van der Waarde for their very helpful comments and suggestions on this chapter.

References

ACIA. (2001). *Short and long term outlook for information architects*. Retrieved 9/7/2001 from the World Wide Web: argus-acia.com/iask/survey010907.html.

Adams, A. (1994). Usability and information design. In H. Zwaga, T. Boersema, & H. Hoonhout (Eds.), *Visual information for everyday use* (pp. 3-20). London: Taylor & Francis.

Advance for Design. (2001). *What is experience design?* Retrieved 8/31/2001 from the World Wide Web: advance.aiga.org/expdesign.

Albers, M., & Lisberg, B. (2000). Information design: A bibliography. *Technical Communication, 47*(2), 170-176.

Arnheim, R. (1969). *Visual thinking*. Berkeley: University of California Press.

Barass, X. (1996). *Auditory information design*. Retrieved 7/30/2001 from the World Wide Web: viswiz.gmd.de/~barrass/thesis.

Card, S., Mackinlay, J., & Shneiderman, B. (1999). *Readings in information visualization: Using vision to think*. San Francisco: Morgan Kaufmann.

Cloniger, C. (n.d.). *Usability experts are from Mars, Graphic designers are from Venus*. Retrieved 8/31/2001 from the World Wide Web: www.alistapart.com/stories/marsvenus.

Cohen, S. J. (n.d.). *The curse of information design*. Retrieved 8/31/2001 from the World Wide Web: www.alistapart.com/stories/curse.

Computer Literacy Bookshops. (1997). Edward R. Tufte—Computer literacy bookshops interview. Retrieved 4/18/1997 from the World Wide Web: www.clbooks.com/nbb/tufte.html

Cooper, A. (1995). *About face: The essentials of user interface design.* Foster City, CA: IDG Books Worldwide.

Cotton, B., & Oliver, R. (1997). *Understanding hypermedia* (2nd ed.). London: Phaidon Press.

Davis, M. (2001). Shaping the future of our profession. *Technical Communication, 48*(2), 139-144.

Dillon, A. (2001). *The state of the profession.* Retrieved 8/31/2001 from the World Wide Web: www.slis.indiana.edu/adillon/IA5.html

Dondis, D. (1973). *A primer of visual literacy.* Cambridge, MA: MIT Press.

Easterby, R., & Zwaga, H. (1984). *Information design: The design and evaluation of signs and printed material.* Chichester, UK: Wiley.

Felker, D., Pickering, F., Charrow, V., Holland, V. M., & Redish, J. (1981). *Guidelines for document design.* Washington, DC: American Institutes for Research.

Garrett, J. J. (2001). Response in "What's in a name?" *Design Matters, 5*(2), 3. Online at: www.stcsig.org/id/newsletter.html

Hamilton, E. (1970). *Graphic design for the computer age.* New York: Van Nostrand Reinhold.

Harris, R. (1996). *Information graphics: A comprehensive reference.* Atlanta, GA: Management Graphics.

Hayhoe, G. (2000). What do technical communicators need to know? *Technical Communication, 47*(2), 151-153.

Horn, R. (1998). *Visual language: Global communication for the 21st century.* Bainbridge Island, WA: MacroVU, Inc.

Horn, R. (1999). Information design. In R. Jacobson (Ed.), *Information design* (pp. 15-33). Cambridge, MA: MIT Press.

Hunter. L. (2000/2001). Document design: Complex evolution. An interview with Karen Schriver. *Document design, 2*(1), 28-33.

Information Design Journal. (1979). *1*(1).

Jacobson, R. (Ed.). (1999). *Information design.* Cambridge, MA: MIT Press.

Kinneir, J. (1984). The practical and graphic problems of road sign design. In R. Easterby & H. Zwaga, *Information design: The design and evaluation of signs and printed material* (pp. 341-350). Chichester, England: Wiley.

Kostelnick, C., & Roberts, D. (1998). *Designing visual language.* Needham, MA: Allyn & Bacon.

Laurel, B. (1993). *Computers as theatre.* Reading, MA: Addison-Wesley.

Laurel, B. (Ed.). (1998). *The art of human-computer interface design.* Reading, MA: Addison-Wesley.

Mazur, B. (1999). Our roots...and our future. *Design Matters, 3*(3), 2.

Mead, J. (1998). Measuring the value added by technical documentation: A review of research and practice. *Technical Communication, 45*(3), 353-379.

Meyer, E. (1997). *Designing infographics*. Indianapolis, IN: Hayden Books.

Mijksenaar, P. (1997). *Visual function: An introduction to information design*. New York: Princeton Architectural Press.

Mok, C. (1996). *Designing business*. San Jose, CA: Adobe Press.

Morville, P. (2000, 27 July). *Big architect, little architect*. Retrieved 8/31/2001 from the World Wide Web: argus-acia.com/strange_connections/strange004.html.

Mullet, K. & Sano, D. (1995). *Designing visual interfaces: Communication oriented techniques*. Englewood Cliffs, NJ: Prentice Hall.

Nielsen, J. (n.d.). *Why this site has almost no graphics*. Retrieved 9/5/2001 from the World Wide Web: www.useit.com/about/nographics.html.

Nijhuis, W., & Boersema, T. (1994). Cooperation between graphic designers and applied behavioral researchers. In H. Zwaga, T. Boersema, & H. Hoonhout (Eds.), *Visual information for everyday use* (pp. 21-42). London: Taylor & Francis.

Passini, R. (1999). Sign-posting information design. In R. Jacobson (Ed.), *Information design* (pp. 83-98). Cambridge, MA: MIT Press.

Pettersson, R. (1998, March). What is information design? In *Proceedings, Vision Plus 4*, Pittsburgh, PA, 58-74.

Raskin, J. (1999). Presenting information. In R. Jacobson, (Ed.), *Information design* (pp. 341-348). Cambridge, MA: MIT Press.

Redish, J. (2000). What is information design? *Technical Communication*, 47(2), 163-166.

Rosenfeld, L. (23 Aug 2001). *Future directions for IA*. Retrieved 9/5/2001 from the World Wide Web: www.louisrosenfeld.com/home/bloug_archive/005.html.

Rosenfeld, L., & McMullin, J. (2001). *IA Areas of Practice*. Retrieved 9/19/2001 from the World Wide Web: www.louisrosenfeld.com/images/bloug/010725b.gif.

Rosenfeld, L., & Morville, P. (1998). *Information architecture for the World Wide Web*. Sebastopol, CA: O'Reilly & Associations.

Schriver, K. (1997). *Dynamics in document design*. New York: Wiley.

Schriver, K. (2001). Response in "What's in a name?" *Design Matters*, 5(2), 8.

Shedroff, N. (1999). Information interaction design: A unified field theory of design. In R. Jacobson (Ed.), *Information design* (pp. 267-292). Cambridge, MA: MIT Press.

Shedroff, N. (2001a). *Experience design 1*. Indianapolis, IN: New Riders.

Shedroff, N. (2001b). Response in "What's in a name?" *Design Matters*, 5(2), 6.

Shneiderman, B. (1997). *Designing the user interface: Strategies for human-computer interaction* (3rd ed.). Reading, MA: Addison-Wesley.

Sless, D. (1998). Building the bridge across years and disciplines. *Information Design Journal*, 9(1), 3-10.

Sless, D. (1999). "The Future of Information Design?" *Design Matters*, 3(3), 6.

Sless, D. (2001). "What's in a name?" *Design Matters*, 5(2), 6.

Spender, D. (1995). *Nattering on the net: Women, power, and cyberspace.* North Melbourne, Vic: Spinifex Press.

Spiekermann, E. (1995). *Erik Spiekermann on information design: What is it? Who needs it?* Retrieved from the World Wide Web: www.metadesign.com/metaculture/articles/info.htm.

Stiff, P. (1995). Public graphics: the Lunteren symposium. *Information Design Journal, 8*(1), 64-71.

Stiff, P. (1999). Foreword. In H. Zwaga, T. Boersema, & H. Hoonhout (Eds.), *Visual information for everyday use* (pp. xi-xii). London: Taylor & Francis.

Taylor, C. (4/27/2000). Clarity, art and news-graphics (InfoDesign-Café post). Retrieved 8/31/01 from the World Wide Web: list.design-inst.nl/pipermail/infodesign-cafe/2000-April/000986.html

Taylor, C. (2000). Information design: A European perspective. *Technical Communication, 47*(2), 167-168.

Thwaites, H. (1999). Visual design in three dimensions. In R. Jacobson, (Ed.), *Information design* (pp. 222-245). Cambridge, MA: MIT Press.

Tufte, E. (1983). *The visual display of quantitative information.* Cheshire, CT: Graphics Press.

Tufte, E. (1989). *Envisioning information.* Cheshire, CT: Graphics Press.

Tufte, E. (1997). *Visual explanations: Images and quantities, evidence and narrative.* Cheshire, CT: Graphics Press.

Waller, R. (1996). The origins of the IDA. Retrieved 8/26/2001 from the World Wide Web: www.informationdesign.org/ida/origins.html.

Westendorp, P., & van der Waarde, K. (2000/2001). *Information Design Journal, 10*(1), 1-3.

White, J. (1982). *Editing by design: A guide to effective word and picture communication for editors and designers* (2nd ed.). New York and London: R. R. Bowker.

Wildbur, P. (1989). *Information graphics: A survey of typographic, diagrammatic and cartographic communication.* New York: Van Nostrand Reinhold

Wodtke, C. (27 Aug 2001) *State of the profession.* Retrieved 9/6/2001 from the World Wide Web: www.eleganthack.com/blog/archives/00000135.html

Wurman, R. (1989). *Information anxiety.* New York: Bantam.

Wurman, R. (1996). *Information architects.* New York: Graphics Press.

Wurman, R. (3 Apr 1998). Information architects / TEDX. (InfoDesign post). Retrieved 8/31/01 from the World Wide Web: list.design-inst.nl/pipermail/infodesign/1998-April/000056.html

Wurman, R. (1999). *Understanding USA.* Newport, RI: TED Conferences Inc.

Wurman, R. (2000). *Information anxiety 2.* Indianapolis, IN: Que.

Wurman, R. (2001). Response in "What's in a name?" *Design Matters, 5*(2), 9.

Zwaga, H., Boersma. T., & Hoonhout, H. (1999). *Visual information for everyday use: Design and research perspectives.* London, UK, and Philadelphia, PA: Taylor & Francis.

Endnotes

1 See www.infomap.com.

2 Karel van der Waarde, editor of the *Information Design Journal*, notes that Neurath's major work actually started in 1924 when he became director of the Gesellschafts- und Wirtschaftsmuseum in Vienna.

3 For a comprehensive resource, see Harris, (1996). For another history, see Meyer (1997). For case studies, see Wildbur, (1989).

4 See www.id.iit.edu/visiblelanguage.

5 Indeed, the primary difference in these types of products is likely to depend on whether one's background is primarily graphic design or technical writing or some variant.

6 Mijksenaar reports that initial capitals were reinstated in Schiphol in 1993.

7 Unfortunately, the Easterby and Zwaga volume is out of print, making it hard to access early work in the field. However it may be found in libraries with strong design or communication holdings.

8 The Medical Research Council, www.mrc.ac.uk.

9 Waller does note that the design firm Pentagram used the term in a book of their work published "at about the same time."

10 Curiously, Robert Horn credits the "British Information Design Society" as the group that "invented and popularized the term *information design*" (1999, p. 21).

11 A number of useful back issues are available at www.informationdesign.org/ida/oldIDeAs.html.

12 See www.iiid.net for more about the IIID.

13 See the later section on information architecture and information design for a discussion on Wurman's choice of the term information architecture.

14 See www.InformationDesign.org for subscription information.

15 See www.stcsig.org/id.

16 STC currently has 25,000 members. See www.stc.org/aboutus.htm

17 See www.InformationDesign.org/ida.

18 See www.benjamins.nl/idj. Note that Benjamins also publishes *Document Design*; let.kub.nl/docdes.

19 Most often in these discussions, "profession" is viewed specifically as a more rigid set of rules, standards, and education (e.g., law or medicine).

20 Schriver's discussion of this in *Dynamics in Document Design* offers interesting parallels; see pp. 53-71 in particular.

21 See Albers and Lisberg (2000) for a comprehensive bibliography.

22 The IDJ is printing a review of *Understanding USA* in issue 10(2).

23 A film from 1976 critical of the U.S. television news industry; the main character, a news anchor, encourages his audience to open their windows and yell "I'm mad as hell and I'm not going to take it anymore."

24 Described humorously by the Abraham Maslow quote: "If the only tool you have is a hammer, you will see every problem as a nail."

25 See *Understanding Hypermedia 2000* for both an extensive timeline (combining events in telecommunications, television and video, radio and records, photography and film, and print and publishing as the precursors to hypermedia) and an exploration of what hypermedia might mean as a communications medium (Cotton & Oliver, 1997).

26 On August 21, 2001, as part of the Chairman's Address at the Progress and Freedom Foundation's Aspen Summit and subsequently televised on C-SPAN.

27 Experience design is a new field with its own historical roots; Nathan Shedroff cites John Dewey's *Experience and Education* (1938) in *Experience Design 1* (2001a).

28 See June/July 2001 discussions of the InfoDesign Café.

29 I've admired Richard Saul Wurman and Nathan Shedroff immensely since I read their early works. But with books like *Understanding USA* and *Experience Design 1*, I'm left with the troubling thought that I am no longer hip enough to "get" the design.

30 See Whitney Quesenbery's chapter in this volume.

31 The title of a session I presented at the 1999 STC Annual Conference.

32 I am less sure that the skills that might enable one to get so many experts coordinated is the same person who manages the budget and time schedule. I am tempted to have the management and budgetary tasks live in the role of the project manager, whose film counterpart is the producer. Obviously this depends on the scale of the project.

33 See www.wordfocus.com/word-act-blindmen.html.

2

Physical, Cognitive, and Affective: A Three-Part Framework for Information Design

Saul Carliner
City University of Hong Kong

This is an exciting time for technical communicators. We're moving from a focus on the tools used to produce content, like help authoring tools and desktop publishing programs, to one on the content itself (Carliner, 2000). According to JoAnn Hackos (2000), sponsors of our work now expect us add value through innovative approaches to information because we have achieved the cost savings promised by automating the production processes.

In some cases, innovative approaches involve experimenting with new ways of presenting content. As many of you readers are already aware, we're embedding information into the user interface (called *embedded user assistance* and *electronic performance support*; Marion 1997, 1998) and developing knowledge management systems that capture, store, transform, and disseminate the information that's crucial to our businesses; Thurman, 1999). In some cases, we are even designing and developing the interaction between computers and users (an activity called *interaction design*; Hewett, 1996).

In some other cases, innovative approaches involve developing a single base of content that can be used in many ways. For example, some organizations print books and display information online from the same source file. Others use a single source file for publishing very different versions of the content, such as one version intended for a training course and another intended as a user's guide (Kostur, 1999).

Design is an essential ingredient to the success of all of these efforts. For example, to develop an online interaction, a technical communicator must not only write the message presented to users, but must first predict users' goals, moods, and motivations, and gear the message accordingly. If several different types of users encounter the same content, then the communicator must also sense this difference and display a message that's tailored not only to the context and mood, but to the type of user.

Similarly, to develop a single base of content that can be used in a variety of systems and contexts, technical communicators must identify all potential uses of information, then prepare designs that accommodate all of these different uses (Hackos, 1999). In some cases, technical communicators must coordinate the designs for several related communication products because they will be based on the same source of content.

This concept of design broadens the role of technical communicators beyond the traditional boundaries of writing and page design. Up-front efforts are far more complex than defining the audience and purpose of the communication product. In most cases, a communication product has several audiences and purposes, and is developed in conjunction with related documentation, and marketing and training materials. Budgets, schedules, existing technology infrastructures, and corporate cultures place boundaries on the solutions available to design challenges. Because of this complexity, some organizations tackle design in two phases.

This broad focus also taxes our existing approaches to document design, and serves as the subject of this article. Specifically, this chapter first explores limitations with the prevailing concept of document design. Next, it offers a definition of information design—a framework meant to broaden the popular perspective on design in our field. The chapter then describes in detail the three types of design activities involved in technical communication: physical design, cognitive design, and affective. Last, this chapter suggests the strengths and limitations of this framework.

Limitations with the prevailing concept of document design

Since the 1980s, technical communication has been guided by a concept called *document design*, which was introduced by Felker, Pickering, Charrow, Holland, and Redish (1981) in their book, *Guidelines for Document Designers*.

According to Felker et al., document design is:

> a field concerned with creating texts (broadly defined) that integrate words and pictures in ways that help people to achieve their specific goals for using texts at home, school, or work. (Schriver, 1997).

According to Redish (2000), document design has two dimensions:

1. The overall process of developing a successful document
2. The way the information is presented on the page or screen (layout, typography, color, and so forth)

Although the definitions accommodate both, practice of technical communication focuses almost exclusively on the second dimension. For example, consider this advertisement for a session for technical communicators on designing web pages that was held at a recent conference:

> The magical formula for designing successful web pages is:
>
> content + writing style + layout = information design… (www.uie.com)

Where are the readers' goals in this magical formula? Where is the process of selecting the content?

This focus on text and appearance is not an isolated one. Consider informal comments by many technical communication students who study document design. They want to learn how to design pages, failing to recognize that a well-designed page only addresses the surface—and does not compensate for an unusable product design or an incompletely thought-through technical concept. Too often, designers begin software documentation assignments by choosing a help authoring system. Only afterwards do they consider the content the users need "help" with.

Somehow, the practice of design as improving the appearance of pages and screens has replaced the concept of design as problem solving, even though published definitions of document design suggest otherwise. Perhaps that's because the source material is primarily a series of guidelines of dos and don'ts for technical communication products—a cookbook of sorts. Although few read the source material any more (the original *Guidelines* are out of print), technical communicators still focus on discrete issues, such as the most appropriate font color and size, and the most usable arrangement of information on the screen, as if a single answer fit every context.

Even when considering document design as the overall process of developing a successful document, that process has been drastically altered since the 1980s. For example, the introduction of desktop publishing in the 1980s significantly shortened the production cycle, but added technology issues that limited some design options and often moved responsibility for copyediting and producing graphics and camera-ready copy to the technical communicator. Similarly, the emergence of online help and web sites expanded our choice of media, and the number of media in which we needed to comfortably communicate. We had to learn how to write for presentation online and design screens. Although the guidelines for document design offered direction, they did not always transfer to the new medium.

Recognizing this limitation, other approaches have been offered to supplement the concept of document design. Many of these broader approaches emerge from practice or studies of practice:

- **Writing as a problem-solving activity.** That is, technical and professional writing are intended to communicate for a specific purpose (Flower, 1989).
- **Task-oriented writing.** That is, content should be structured to assist people with specific tasks that they need to perform, rather than as documentation of features and functions, from which users then must infer how to perform the desired tasks.
- **Minimalism.** That is, users can become productive quickly with a limited amount of instruction, but technical communicators have to carefully choose which information to provide—and which to leave out (Carroll, 1990).
- **User-centered design.** That is, following certain principles of design can make information easier to use and requires less documentation of products and their interfaces (Norman, 1990). Technical communicators' interest in these design principles rose as information for consumers and novice users became a significant part of the work.
- **Human performance technology.** That is, technical communication activities attempt to assist individuals and groups in achieving goals. Emerging from the field of instructional design, performance technology (also called performance-centered design) involves more than imparting knowledge and skills. It requires that designers make sure users have the proper resources and promote supportive attitudes and cultures so that performers can accomplish these goals (Stolovich & Keeps, 1999, Bowie, 1996).

- **Bottom-line impact**, which proposes that communication products should be linked to the organizational goals and financial performance of the organizations that commission them (Robinson & Robinson, 1989).
- **Process-maturity**, which proposes that the quality of products is related to the maturity of the process that developed them (Hackos, 1994).

Each approach addresses a different limitation of document design, but none comprehensively incorporates the strengths of other approaches or describes the relationship of each issue to one another.

Perhaps these limitations of document design have spurred some technical communicators to embrace a notion called information design. Indeed, some call themselves information designers or information architects. Privately, these people who identify themselves as information designers might say, "I'm not a technical stenographer," "I do more than wordsmith programming specifications" or "make pages and screens look good." Like architects of buildings, information designers look at the bigger picture: What problem is the client trying to solve, what can I bring into play to address the problem, and how does this solution support the larger business situation.

But what is information design? The Vienna-based International Institute of Information Design (1997) admits that information design

> can be hard to define, because it is an interdisciplinary approach which combines skills in graphic design, writing and editing, illustration, and human factors. Information designers seek to combine skills in these fields to make complex information easier to understand.

Without a precise definition that distinguishes information design from document design, and without offering an alternative to document design, information design looks like nothing more than a marketing ploy.

Recent professional and scholarly activities have attempted to make information a topic of academic discussion. Most academic discussion has occurred in the circles of architects and graphic designers. For example, architect Richard Saul Wurman's (1996) *Information Architects* presents case studies of practical information that has been designed for effective, easy use. The edited collection *Information Design* (Jacobson, 1999) primarily applies graphic and architectural design theory to the design of communication products. Neither of these books really presents a practical definition of information design.

A definition of information design

Before offering a more comprehensive definition of information design, a consideration of design in general might be appropriate. According to Rowland (1993):

> Some argue that a science of design is possible and represents an important goal. Cross, reporting on a number of studies in design, argues that design is quite different from science. While scientists focus on the problem, on discovering the rule that is operating, designers focus on the solution, on achieving the desired result. (p. 81)

He concludes that design is ultimately personal choice based on a subjective sense of what is "right."

In other words, design is a problem-solving discipline. It considers more than the appearance of the designed product, but also the underlying structure of the solution and its anticipated reception by users. Because design is focused on solving problems, a design theory must provide more than a series of guidelines about discrete characteristics of the solution; it must focus designers on identifying the problem and supplying a framework for identifying and considering the interrelated issues that must be addressed in a solution. Design must also help designers develop their instincts for choosing "right" solutions.

If information design primarily focuses on issues of appearance and text, it is neither distinct from document design nor solves the problem of the limited focus of document design in most current practice and research.

Information design must therefore have a broader focus, one that encompasses not only graphics and text, and reader goals, but also the goals of the sponsor who commissioned the text. Therefore, information design may be better defined as "preparing communication products so that they achieve the performance objectives established for them." This involves:

- Analyzing communication problems.
- Establishing performance objectives that, when achieved, address these objectives.
- Developing a blueprint for a communication effort to address those objectives.
- Developing the components of the planned communication effort solution.
- Evaluating the ultimate effectiveness of the effort.

Some of the terms in this definition have specific meanings.

- *Performance objectives* are observable, measurable tasks and business goals that users should be able to perform, the conditions for doing those tasks, and the level of acceptable work (Mager, 1997a).
- *A blueprint* is a detailed design plan for a document that indicates not only the content to be presented, but the extent and format of the presentation (Kostur, 1999).

Inherent in this definition of information design is the Analysis, Design, Development, Implementation, and Evaluation (ADDIE) model that is widely used in instructional design (Gustafson, 1991) and similar to ones used in software engineering.

A model of information design

This new model approaches information design on three levels. This model is adapted from the three levels that theorists in education and instructional design consider when designing courses (Dick & Carey, 1990).

- **Physical**, the ability to find information
- **Cognitive** (intellectual), the ability to understand information
- **Affective** (emotional), the ability to feel comfortable with the presentation of the information (comfort with the information itself might not be possible, depending on the message)

In the next three sections, I address each of three levels in detail, explaining what it is, how it relates to current research and discussions of technical communication, and naming the elements of design and some of the key design issues addressed. Table 2.1 summarizes this model.

The physical level: Helping users find information
The first type of design is the physical. From the users' perspective, good physical design lets them find information of interest easily. For example, if users seek information about the DOS command named "dir," can they easily locate that information? The physical level also concerns the general appearance of information.

Table 2.1. Physical, Cognitive, and Affective—A three-part model of design for technical communication products.

Physical design: helping users find information
Issues
- Page and screen design
- Retrievability
- Media selection
- Production
- Basic technical writing and editing

Cognitive design: helping users understand information
Process
1. Analyzing needs
2. Setting goals
3. Choosing the form
4. Preparing the design
5. Setting the guidelines

Issues
- Applying principles of cognitive psychology
- Applying design theories such as human performance technology and minimalism
- Addressing potential information overload
- Modularizing information
- Designing within constraints

Affective design: motivating users to perform
Issues
- Attention
- Motivation
- Change management Cross-cultural communication
- Language
- Social and political impact
- Legal and ethical issues
- Client service
- Methodologies for understanding communication issues

The design issues associated with physical level are typically those associated with the traditional practice of document design, including:

- **Page and screen design,** which refers to placing information on a page or screen so readers can easily locate it. Elements of page and screen design include:
  o *Layout,* or placement of information
  o *White space*
  o *Headings,* which readers can scan to find information
  o *Type,* which can help or hinder the search and reading process
  o *Graphical devices,* which can call readers' attention to key elements of information
- **Retrievability aids,** which refer to those devices that help readers locate information in a document. These aids include tables of contents, site maps, indexes, links, running headers and footers, and tabs.
- **Media selection,** which refers to choosing the appropriate means of physically delivering the information to users. Information can be delivered in print, online, through video or audiotape, or a live connection.
- **Production,** which refers to the process of preparing a communication product for duplication and distribution to its intended readers. Elements of production include:
  o Copymarking text to conform to style guidelines
  o Preparation of graphics and other media elements
  o Integration of text, graphics, and other media elements into a master copy
  o Preparation of materials for printing, CD duplication, etc.
  o Packaging
  o Software or tools for preparing text, graphics, and other media elements
- **Basic technical writing and editing,** which refers to the skill of preparing text that conforms to a generally acceptable style. Examples of text preparation include: composing instructions in the active voice, and preparing scientific articles according to the guidelines for authors of the intended publication.

Physical design plays an important role in the overall design of communication products. But physical design elements are only cosmetic if information designers do not consider them as part of a larger, goal-oriented framework. That is where the intellectual level plays a key role.

The cognitive level: Helping users understand information

The second level of design is the cognitive, or intellectual level. That is, once readers find information, can they understand it and make use of it? For example, once readers find the description of the "dir" command, can they follow the instructions and actually use it?

People who call themselves information designers typically address cognitive design issues. Richard Saul Wurman (1989), for example, called this "the understanding business."

Cognitive design primarily focuses on the design process: adequately defining the users' performance goals and preparing a solution that addresses them. Mirel (1998) suggested the need for technical communicators to shape instructions around the problems that people actually experience in work contexts and adopting problem-based instruction. The instructional systems design (ISD), which is widely used by instructional designers (Foshay, 1997; Wedman & Tessmer, 1993; Zemke & Lee, 1987), offers a framework for such a model.

Although cited as a model, ISD is actually just a process for designing instructional programs so that they achieve key learning objectives (Dick & Carey, 1990). In this process, however, ISD provides a structure for considering the countless issues that must be addressed when analyzing and defining the problem, and devising its solution. For example, ISD addresses issues such as usability (in its call for clear objectives and formative evaluation), adjusting to the styles of different levels of learners, and re-using or adapting existing courses (if they meet the stated objectives of the proposed course).

In the sense that it follows a sequence of activities to identify the users and their performance goals, and provides them with the right information at the right time, cognitive design can be seen as a process. Specifically, the design process consists of the following activities:

- **Analyzing needs,** to assess the following:
  - o The business need underlying the request for a communication product
  - o The situations that drive users to seek information or assistance, called scenarios or use cases (Nurminen & Karppinen, 2000). These scenarios describe the type of need motivating the search, and the goals that readers hope to accomplish to resolve the situation.
  - o The tasks that readers must successfully complete, listed in the likely order in which they'll be performed, when achieving the goals that drove them to search for assistance.
  - o The motivations, needs, and experience of the key groups of anticipated readers.

- o Constraints affecting the ability to successfully produce the communication product, such as a must-meet completion date, not-to-exceed limit of the budget, expectations of quality, quirks in the corporate culture of the requesting or producing organization, and personalities involved in the project (Hackos, 1994; Robinson & Robinson, 1989; Rossett, 1987).
- **Setting goals for the project,** which includes:
- o Establishing observable and measurable business and content objectives. Business objectives, for example, define how the communication product directly or indirectly contributes to the revenues of the sponsoring organization, contains their expenses, or assists in meeting corporate, industry, or government regulations placed on the sponsoring organization (Carliner, 1998). Content objectives describe, in observable and measurable terms, the main and supporting tasks that readers should be able to perform after using the communication product (Hackos, 1994; Mager, 1997a).
- o Planning a comprehensive evaluation to ensure that the communication product meets its objectives. Mager (1997a) noted that planning an evaluation first helps designers visualize the end result, and then helps them design for success. A comprehensive evaluation employs several types of assessments, including assessments of reader satisfaction (such as a Reader's Comment Form), usability, and business performance (Carliner, 1997).
- **Choosing the form of the communication product,** which involves choosing:
- o The genre of communication product, such as a user's guide, help system, tutorial, or job aid. In the field of rhetoric, a *genre* may be defined as a language system that contains "typified responses to events that recur over time and across space and that emerge in the social context of communication practices" (Berkenhooter & Huckin, cited in Killingsworth, 1996, p. 107). Each given genre has a group of characteristics associated with it. To be fully considered as an example of a specific type of genre, a communication product must contain all of those associated characteristics (Foss, 1995). Readers bring a set of expectations to a communication product based on its form. For example, they expect user's guides to provide: step-by-step procedures for most common tasks, a table of contents that immediately directs them to the task, and illustrations. They also expect that the step-by-step procedures will be written in the imperative mood.

> o The communications medium for delivering the product to users, an issue also considered as part of physical access.
- **Preparing the design of the communication product,** which is similar to an architect preparing blueprints for a building. As architects develop a floor plan, elevation drawings, and schematics to describe exactly how a building will look and how its internal systems function when completed, so information designers prepare:
  > o The structure of the communication product, as represented by an *information map*, a diagram that shows the structure of information in a communication product, much as a floor plan shows the location of rooms in a building and a flow chart shows the sequence of activities in a program. This representation of the structure should also include the front and back matter of the communication product.

Note that technical communicators have traditionally prepared outlines to represent the structure of proposed communication products. Anecdotal experience suggests that sponsors and users can more easily follow information maps.

- A sample section of the communication product, to formalize the initial design and serve as a model for subsequent discussion and evaluation. A sample section is similar to the model that many architects prepare for proposed buildings.
- Detailed storyboards or thumbnails detailing—screen-by-screen or page-by-page—which objectives are covered, how the content associated with those objectives will be presented (graphically, through video or audio, or in text) and which production and programming elements are needed to realize these plans. Storyboards are like the elevation drawings that architects prepare for buildings, which show how a building looks from a variety of perspectives such as the front and back.
- Specifications of designs for recurring types of pages or screens, such as pages containing procedures or introductions to new sections. Specifications would cover such issues as layout, margins, type, and required and recurring graphical images.

Information designers prepare separate blueprints for each communication product, even if those products are related to one another.

- **Setting the guidelines:**
  - o Product guidelines, including the editorial guidelines (style guide, dictionary, and a list of any exceptions) production specifications (such packaging for online products and printing specifications), and technical specifications for authoring (such as the configuration of the authoring workstations). Product guidelines for communication products are like the specifications for architectural plans, which identify the types of materials to be used in a building such as the type of pipes and moldings.
  - o Project guidelines, including the schedule, budget, and staff. Waiting until after the communication product is designed before providing a detailed project schedule and budget for final completion ensures more accurate budgeting. Projects estimated before such plans have been developed can be off of estimates by as much as 20 to 25 percent. Because estimates are developed after design plans are finalized and based on more thorough information, estimates provided at this point in a project are typically within 2 to 3 percent of estimates (Foshay, 1997). Project guidelines are like the construction plan for a proposed building.

Although essential for the communication product, the design process is only one part of cognitive design. The other part involves defining the intellectual capacities and needs of users, and crafting an appropriate solution to meet these needs. Designers can consider these issues from a number of perspectives.

Cognitive psychology explores the ways in which the human brain processes information. By applying this knowledge, information designers can "pre-digest" information to minimize the amount of cognitive processing needed, just as LactAid pre-digests lactose so that people with a lactose intolerance can eat and drink dairy products.

Design theories, such as human performance technology, minimalism, user-centered design, and constructivism, also inspire cognitive design. Although each cognitive design theory has a unique definition, each of their goals is remarkably similar: providing users with the most appropriate information, at the exact time and place they need it.

Because users are routinely exposed to more messages from the media than they can effectively assimilate, information designers must also consider information overload. Some recommended solutions to this problem involve changes in physical design, such as communicating through visuals rather than words. Other proposed solutions involve changes in cognitive design, such as creatively structuring information and mpracticing "benevolent censorship," that is, removing less essential information (Wurman, 1989).

Another consideration for cognitive design is reusing or modularizing information. In some instances, organizations can effectively reuse information, either in part or in whole. For example, television news organizations often use the same story in several different newscasts, editing the story to fit the length and slant of the new program. In other instances, organizations might mix and match information, such as creating a new manual for Model B by adapting the existing manual for Model A—with the fewest necessary modifications. Or, an organization might let readers mix and match modules of information so that they can effectively customize communication products based on their specific needs. To design information once and use it again in a variety of contexts, with few or no changes, requires careful design of individual units of information and thorough consideration of all possible uses.

A final issue of cognitive design is reconciling design plans with business realities. Most technical communication projects are constrained by budgets, schedules, editorial guidelines, or some combination of these. According to Schriver (1999), expert document designers can deftly coordinate design and business priorities, yet few design models actually incorporate all or even most of these issues.

The affective level: Motivating users to perform

The last level of design is the affective, or designing the communication product for its optimum emotional impact. That is, if users can find the information they need and understand it, is it written in such a way that users will want to use it and perform the intended tasks? For example, if users can find and read information about the "dir" command, will they want to use it, after all?

The issues associated with affective design typically fall into an emerging realm called communication design. Following are the elements of affective design.

Attention. Before users can perform the tasks describe in communication products, they need to feel compelled to read about them. As the fields of advertising, book publishing, and training, among others, have learned that getting positive attention is essential to the success of a product, so technical communicators are learning the same lesson.

Motivation. After attracting readers' attention, technical communicators must motivate readers' to use information in the communication product. For example, "must" people use information to perform their jobs better or is the information "nice to know" but of little practical value? Users' attitudes vary with their motivation so we must address attitudes to successfully transfer information about tasks.

Change Management. This is also called transfer of technology. The technology described in technical communication products often has the potential to change the way users work or live. But how prepared are users for the change? How can the communication product address the anxiety and apprehension that are often byproducts of change?

Language (also called word choice). Although the grammar of a language or the lexicon of a technical discipline often dictates the choice of word used to express an idea, in many other instances, this choice derives from the conscious choice to project a particular image. For example, some people insist on using technical language to gain credibility with technical readers, even if simpler terms might express the same concepts. In other instances, words that seem innocuous to the communicator carry strong meanings for the audience. The Plain Language movement investigates the use of language and its impact on readers (Mazur, 2000), as does the field of rhetoric, which many cite as the primary discipline from which technical communication emerges.

Cross-Cultural Communication. Communication products are often published by people who have little or nothing in common with the intended users. The issue of communicating across national cultures and languages has been well documented (Hoft, 1995), but one need not leave the neighborhood before finding new cultures. Other types of cultural differences that technical communicators need to address includes occupational cultures (Trice & Beyer, 1993), urban versus suburban cultures, and socioeconomic cultures within the same national and language groups.

Social and Political Impact. Even the most seemingly benign message can carry with it a variety of social and political implications. For example, how does the message affirm or undermine the existing political structure of the organization that supports its? Researchers and theorists in technical communication focus much of their energy on this particular issue, though it receives little attention by practicing professionals.

Legal and Ethical Issues. The process of communicating technical information often encompasses a variety of legal and ethical issues, such as copyright, privacy, and implied and stated promises. For example, when preparing to publish a marketing database, Lotus encountered a public angry about the potential invasion of privacy (Gurak, 1999). Intel encountered a similar challenge when the public learned that the Pentium III chip could send a computer's serial number without users' knowledge.

Client Service. Although most literature on document design emphasizes the primary role of the user (Schriver, 1997; Dumas & Redish, 1999), that technical communicators write for users is a "myth" (Sakson, 1996). Rather, technical communicators prepare works for hire that are commissioned by sponsors, such as the Development or Marketing organization inside an organization or an external client.

Therefore, one of their primary concerns for technical communicators should be meeting the needs of these sponsors who commission our work. In many instances, meeting sponsors' needs involves work beyond the scope of writing and editing. It involves building trust and confidence in the sponsor. This can be done by anticipating the impact of decisions on individuals and the organization , regularly communicating the status of projects, and following widely understood processes (Fredrickson, 1992; Carliner & Fredrickson, in press).

Methodologies for Understanding Communication Issues. One of the most significant debates within the community of researchers in technical communication is that of research methodology. Researchers debate the role of heuristics and critical research (Charney, 1997). To what extent should communicators trust empirical research and to what extent must they rely on critical thinking (Sauer, 1997)? The study of communication design encourages the exploration of such issues.

Affective design often poses some of the greatest challenges to technical communicators. Technical communicators often seek formulaic approaches to complex issues of communication design. As professionals, we must avoiding giving too much credence to simplistic catalogs of rules like "the five issues to avoid when writing for international audiences." Perhaps it is this tendency that has limited document design from being used in its fullest definition. The complexity of these issues defy simplistic responses.

Rather, they involve anticipating the impact of communication on the intended audiences and the sponsors who commission the work. They involve addressing the negative fallout that might result from too little analysis of the communication product. That is, communication design is a form of "documentation therapy" in which technical communicators diagnose communication and performance issues, and offer reasoned recommendations for resolving them.

Strengths and limitations of this framework

Several strengths characterize the three-part framework of this design model—its physical, cognitive, and affective levels. The first is that it realistically reflects the broad focus of today's technical communicators. Since the 1970s, technical communicators' responsibilities have grown from that of wordsmiths of technical specifications to designers and testers of user interfaces. This model acknowledges that broader role.

A second strength is that the framework incorporates the diverse work of technical communicators into a single frame of reference. Some technical communicators work on desktop publishing, others design interfaces, and still others consult on various aspects of change management. This design model encompasses all of those roles and places them in an appropriate context.

A third strength of this framework is that it addresses many of the issues raised in the academic community of technical communicators, but which are only given passing acknowledgment in the professional community. The design model works to close the gap between those who teach technical communication and those who practice it. Although the framework does not necessarily resolve many of the current debates and incorporates viewpoints that are not shared by all groups within the technical communication community, it does provide a voice for diverse approaches and lets designers determine for themselves how to resolve the differences.

A final strength of this framework is that it incorporates the growing influences of several related fields. For example, in its process orientation, the model displays the influences of software engineering (Hackos, 1994) and instructional design (a field that is also called instructional or educational technology). In its acknowledgment of the physical, cognitive, and affective, the framework acknowledges the influences of educational psychology and instructional design. In its emphasis on measurement and evaluation, the framework acknowledges the influences of adult learning theory and business management.

But the model admittedly suffers from some limitations. The first is that it is prescriptive. That is, the model prescribes the way practice should work; actual practice could substantially vary from the framework as is suggested by studies of the actual use of models of instructional systems design process (Zemke & Lee, 1987; Wedman & Tessmer, 1993).

A second limitation relates to the first: an overlap among the three levels. For example, although they are listed as elements of physical design, writing and substantive editing skills are also elements of cognitive and affective design. Naming an issue and placing it within the context of the framework calls attention to the issue, but does not always adequately describe the full breadth of the issue. In any case, clear distinctions among the different but related issues do not always exist.

Finally, the framework incorporates research and theory, but does not directly emerge from it. That is, it provides a structure for considering many issues addressed by the research and how it relates within the larger context of the everyday world. But the framework itself is the creation of one mind; it did not emerge from the direct, scientific observation or review of working technical communicators.

Notes

I would like to thank George Hayhoe, editor of *Technical Communication*, for his support and encouragement during the development of this chapter. For additional information on possible implications of this framework, see my article by the same name in the Fourth Quarter, 2000 issue of *Technical Communication*.

References

Bowie, J. S. (1996). Information engineering: Using information to drive design. *Intercom, 43*(5), 6-10 & 43.

Carliner, S. (2000). Trends for 2000: Thriving in the boom years. *Intercom, 47*(1), 12-15.

Carliner, S. (1998). Business objectives: A key tool for demonstrating the value of technical communication products and services. *Technical Communication, 45*(3), 380-384.

Carliner, S. (1997). Demonstrating the effectiveness and value of technical communication products and services: A four-level process. *Technical Communication. 44*(3), 252-265.

Carliner, S. (1994). Guest editorial: a practitioner's call for the STC to establish a research agenda. *Technical Communication, 41*(4).

Carliner, S., & Fredrickson, L. (in press). Quality: It's a judgment call. In R. Grice & L. Ridgeway (Eds.), *Improving quality and productivity of communication: Theory and practice.* Washington, DC: Society for Technical Communication.

Carroll, J. (1990). *The Nurenberg funnel: Designing minimalist instruction for practical computer skill.* Cambridge, MA: MIT Press.

Charney, D. (1997). From logocentrism to ethocentrism: Historicizing critiques of writing research. *Technical Communication Quarterly, 7*(1), 9-32.

Dick, W., & Carey, L. (1990). *The Systematic design of instruction* (3rd ed.), New York: HarperCollins.

Dumas, J. S., & Redish, J. C. (1999). *A Practical guide to usability testing.* Intellect: Bristol, England.

Felker, D. B., Pickering, F., Charrow, V. R., Holland, V. M., & Redish, J. C. (1981). *Guidelines for document designers.* Washington, DC: American Institutes for Research.

Flower, L. (1989). *Problem-solving strategies for writing.* San Diego, CA: Harcourt Brace Jovanovich.

Foshay, R. W. (1997, April). *Fourth generation instructional design.* Paper presented at the International Society for Performance Improvement Conference, Anaheim, CA.

Foss, S. (1995.) *Rhetorical criticism: Exploration & practice.* (2nd ed.). Prospect Heights, IL: Waveland Press.

Fredrickson, L. (1992). Quality in technical communication: A definition for the 1990s. *Technical Communication, 39*(3), 394-399.

Gustafson, K. L. (1991). *Survey of instructional development models.* Syracuse, NY: ERIC Clearinghouse on Information Resources.

Gurak, L. (1999). *Persuasion and privacy in cyberspace: The online protests over lotus marketplace and the clipper chip.* New Haven: Yale University Press.

Hackos, J. (2000). Trends for 2000: Moving beyond the cottage. *Intercom, 47*(1), 6-10.

Hackos, J. (1999). Documentation databases. Boston, MA: *1999 Help Technology Conference.* Solutions, Inc. August 26.

Hackos, J. (1994). *Managing your documentation projects.* New York: Wiley.

Hewett, T. (chair). (1996). ACM Special Interest Group on Computer-Human Interaction (SIGCHI) Curriculum Development Group. Curricula for Human-Computer Interaction. www.acm.org/sigchi/cdg/cdg2.html

Hoft, N. (1995). *International technical communication: How to export information about high technology.* New York: Wiley.

Information Design Association. (1997). members.magnet.at/simlinger-iiid/English-2.html

Jacobson, R. (1999). Introduction: Why information design matters. *Information design.* Cambridge: MIT Press.

Killingsworth, M. J. (1996) . Genre knowledge in disciplinary communication: cognition/culture/power. *IEEE Transactions on Professional Communication, 39*(2), 107-8.

Kostelnick, C., & Roberts, D. D. (1998). *Designing visual language: Strategies for professional communicators.* New York: Allyn & Bacon.

Kostur, P. (1999, September). *Developing single-source documentation.* International Professional Communication Conference, New Orleans, LA.

Mager, R. (1997a). *Preparing instructional objectives : A critical tool in the development of effective instruction* (3rd ed.). Atlanta, GA: Center for Effective Performance.

Mager, R. (1997b). *Measuring instructional results* (3rd ed.). Atlanta: Center for Effective Performance.

Marion, C. (1998). EPSS: What does it mean to you? *Intercom, 45*(10), 10-15.

Marion, C. (1997). What is performance-centered design? *Intercom, 44*(6), 9-13.

Mazur, B. (2000). Revising plain language. *Technical Communication, 47*(2), 205-211.

Mirel, B. (1998), "Applied Constructivism" for user documentation: Alternative to conventional task-orientation. *Journal of Business and Technical Communication, 11*(1), 7-49.

Norman, D. (1990). *The design of everyday things.* New York: Doubleday.

Nurminen, M., & Karppinen, A. (2000, May). Use cases as a backbone for document development. 47th STC Annual Conference, Orlando, FL.

Redish, J. (2000). What is information design? *Technical Communication,*
47(2), 163-166.

Robinson, D., & Robinson, J. (1989). *Training for impact.* San Francisco,
CA: Jossey Bass.

Rossett, A. (1987). *Training needs assessment.* Englewood Cliffs, NJ:
Educational Technology Publications.

Rowland, G. (1993). Designing and instructional design. *Educational
Technology Research and Development,* 41(1), 79-91.

Sakson, D. (1996, March), *Keynote presentation.* 1996 Society for Technical
Communication Student Conference, Macon, Georgia.

Sauer, B. (1997). A comment on "Women and feminism in technical
communication: A qualitative analysis of journal articles." *Journal of
Business and Technical Communication,* 13(4), 463.

Schriver, K. (1997). *Dynamics of document design.* New York: Wiley.

Schriver, K. (1999, February). *The evolution of software user assistance from
the perspectives of information designers and users.* Online help conference. Winwriter's, Inc., Seattle, WA.

Stolovich, H., & Keeps, E. (Eds.). (1999). *Handbook of human performance
technology: Improving individual and organizational performance worldwide* (2nd ed.). San Francisco: Jossey-Bass.

Thurman, S. (1999). Knowledge management means career opportunity.
Intercom, 46(7), 18-21.

Trice, H. M., & Beyer, J. M. (1993). *The cultures of work organizations.*
Englewood Cliffs, NJ: Prentice-Hall.

Wedman, J., & Tessmer, M. (1993). Instructional designer's decisions and
priorities: a Survey of design practice. *Performance Improvement
Quarterly,* 6(2), 43-57.

Willis, V. (1990, March.) Presentation to the Atlanta chapter of the International Society for Performance Improvement. Atlanta, GA.

Wurman, R. S. (1996). *Information architects.* New York: Graphics Press.

Wurman, R. S. (1989). *Information anxiety: What to do when information
doesn't tell you what you need to know.* New York: Doubleday.

Zemke, R., & Lee, C. (1987). How long does it take? *Training,* 24(6), 75-80.

3

Collaborative Processes and Politics in Complex Information Design*

David Sless
Communication Research Institute of Australia

Information design has a long history, most of which predates our current preoccupations with software by thousands of years (figure 3.1).

Figure 3.1. 40,000 year old bone fragments believed to be a lunar calendar (Marshack, 1972).

* Adapted from a Keynote address given at the *Co-Designing Conference*, Coventry University, UK, 13 September 2000.

Mindful of that history, I tend to see the integration of information design into software development as just one instance of a continuum of activity. As such, in this chapter I present a view of designing complex information systems within a broader context, which encompasses the collaborative and political nature of information design.

Overall, the processes I describe are those we have found as relevant to designing software documentation as they are to the design for print, or for any other medium. Some aspects of software systems have greatly changed the ways in which we design information, and I will discuss those in turn.

I draw on experiences in information design research at the Communication Research Institute of Australia since 1985. I begin with a brief introduction to the intellectual position from which our work at the Institute proceeds.

Collaboration—CRIA's intellectual position

First, we have been exploring what is often called collaborative design methods or, more recently, user-focused design methods. But collaborative design methods that involve the active participation of the eventual users of a design are not new. Only the terms "collaborative" or "user-focused" are new. The first conference on what was then called "participatory design" was run by the Design Research Society in the UK in 1971. Much of this work grew initially out of engineering, architecture, town planning, and product design. Its application in information design came later. Although many would look at the work reported in those 1971 conference papers and say, "How crude, how primitive," the seeds of our own work at CRIA can be traced to that earlier work. Primarily, in following this tradition we are motivated by a concern for the people who have to suffer the consequences of the designs we create. The very least we can do is meaningfully involve them in the design process.

Second, CRIA takes what is sometimes called a constructionist view of communication—not to be confused with a constructivist view. The constructivist view, which we do not hold, says that our minds construct our social realities; the emphasis is on cognition, perception, and private schemata. The constructionist view, on the other hand, says that we construct our social realities through communication; the emphasis is on dialogue, conversation, and public language. Our philosophical ideas derive in large part from Wittgenstein's *Philosophical Investigations*. Briefly, our interest is very much at the surface of things. We do not believe in, or try to find, deep causation. We work on the things that we can observe, the things that catch our attention. We do not offer any theories of mental functioning. We take little notice of research from other areas that seeks to place both design and communicative activity as something that goes on in people's heads. Our concern is for what goes on between people.

Third, some have thought that, because we are concerned with observable actions, we must be following a behaviorist tradition. There is a subtle but important distinction between action and behavior. We are concerned with what people do—how they act in the world, not the stimuli they respond to. We work within the humanist tradition. Above all else our duty and our role is to respect other people. The material we deal with is the ordinary stuff of everyday life—the things that happen between people and organizations at counters and call centers, on forms and notices—the prosaic, the ordinary, the everyday occurrences that people share. My primary concern as an information designer is to focus on making that information accessible and usable. As part of the humanist tradition, I focus not on the object that I create but on the relationship that I enable.

This, then, is the Institute's major intellectual position, and what we understand by "collaborative process."

Information design at CRIA—The stages

Information designers design such things as forms, timetables, wayfinding systems, documents, labels, websites, and, more recently, entire information systems.

What follows is a generalized picture of what an information design project might look like. Do not take it as prescriptive—I am not suggesting that this is the only way to do it. Any attempt to shed light on a complex process is usually a synthesis, a construction: you do certain things on some projects and other things on others. But as a collaborator in many projects I can identify a number of stages in the process (figure 3.2).

Scoping

Figure 3.3 shows a typical document from a case history showing the tasks undertaken at the scoping stage.

The scoping stage involves about 16 or 17 different investigations. One important investigation is to find out what is wrong with an existing document. We will ask, perhaps, "How many errors do people make when filling in this form you want us to redesign?" Interestingly, managers usually do not know. One government education department assured us there were no errors on their student grant application form—none whatsoever. On further probing we found that the mailroom assistant went through each form that came in, and if it had an error on it she

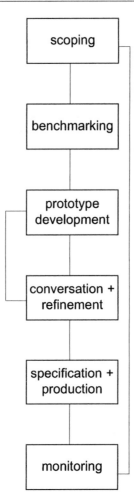

Figure 3.2. Stages in an information design project at CRIA.

returned it to the student for correction, without sending it for further processing. The staff who were assessing the forms thought they were getting error-free returns. This kind of thing is not unusual.

Benchmarking

Benchmarking is essential. Much design work is *re*-design rather than design from scratch. An important part of re-design is to ask: What is the current performance of this design, and what do we want to achieve? Often, in design projects, the urge to redevelop leaves no time to ask those questions; but if we do not know where we currently are and

Stage 1: scoping

In this initial stage we ensure that all factors necessary for the redesign of xxxxxx are known and taken into account.

Our scoping investigations:

- identify and engage with all relevant internal stakeholders. This ensures that internal concerns are identified and managed throughout the project.

- develop an understanding of the history of proposals for changing the xxxxxx. This takes account of sensitivities within the organisation. Where possible, we draw on and make positive use of the accumulated knowledge and ideas within the organisation, so that their prior efforts are acknowledged and, where appropriate, used.

- take account of organisation business plans and strategic directions.

- identify and record organisation business rules relevant to the administration of xxxxxx.

- identify existing data fields for xxxxxx.

- identify regulations and other compliance issues that have to be taken into account in the design of xxxxxx.

- identify the system constraints that have to be taken into account in developing xxxxxx.

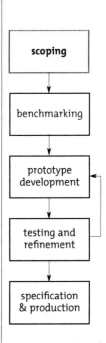

xxxxx systems
© 2001
communication
research institute
of australia
ABN 54 824 399 669

PHONE
61 262 598671
FAX
61 262 598672
E-MAIL
cria@communication.org.au
WEB SITE
www.communication.org.au

page 2

Figure 3.3. Some of the tasks undertaken during the scoping stage.

where we want to go, we will not know when we have arrived. Also, all design activity is generative; that is, it leads to outcomes that none of us foresees, opening things up and providing new possibilities. Benchmarks guide us through the processes and possibilities. And, importantly, benchmarks make good before-and-after politically potent stories: 'When we started, it was like this—but look how great it is now!'

When we conduct benchmarking studies we find out about each of the following criteria. These criteria are more than quantifiable benchmarking. They are the very framework for understanding communication, the fundamental things without which communication and information design will not work.

- The material must be respectful of the people who will use it.
- The material should be attractive, for all sorts of reasons, including the fact that someone actually cares about the material.
- The material has to be usable. CRIA applies strict benchmarking criteria to judge this. Typically, we require that everyone using information should be able to find, understand, and act on at least 80% of what they look for. CRIA provided the guidelines for the Australian government's consumer medicine information regulations, and these minimum usability benchmarks are now built into the guidelines.
- It has to be efficient. Organizations need to know if the money they have spent on a service has added to, or reduced, expenses.
- It has to be physically appropriate. Many objects can be physically difficult to handle. Our research has shown, for example, that for many older people, inserting a document properly into a window envelope can be a major task.
- It has to be socially appropriate. To understand this, I offer an example of the socially inappropriate. Our taxation office refers to you and me not as citizens, nor as taxpayers, but as *clients*, and describes itself as offering a service, much in the way that a lawyer or a prostitute does. But it is a service that I cannot reject, nor can I shop around for an alternative tax office. I do not mind paying tax; but to refer to me as a client in that context is unacceptable; it is a misrepresentation of the relationship. I am not a client of the tax office but a citizen contributing to society through their office. At CRIA, we ask people what relationship they have with their texts and documents.

(CRIA has developed a free Diagnostic Kit which managers can use to test the well being of their organizations against these criteria. You can download it from www.communication.org.au/html/diagnostic.html.)

Developing prototypes

Only when the scoping and benchmarking have been done is a designer in a position to write the design brief. Although the development of prototypes is what most designers and design courses focus on, for us it is one stage in an overall process, albeit an important one, as the prototype development process is informed by the scoping and benchmarking data. But of course these data cannot in themselves create a prototype. This is where the special skills associated with traditional writing and design education come into play. In information design, as in other design areas, technical skill, aesthetic sensibility, and imagination are vital at this stage.

Conversation and refinement

In collaborative work, the challenge is to integrate the prototyping stage of design properly into an overall process. At the conversation and refinement stage, we take a prototype out of the studio and into the world of the people who engage with our designs once they are published or manufactured. Using the benchmark criteria as a guide, we ask people to engage with our design. We watch what they do and join them in conversation to find out what is wrong with our design, so that we can improve it by refinement. It is at this stage that the value of benchmarking becomes obvious; the criteria that we use in the benchmarks are used here.

This, emphatically, is not usability testing; it has little to do with either usability or testing. Usability is too narrow a term to encompass the full range of benchmarking criteria that are employed; and what is happening is not testing, but open-ended conversation or dialogue. We do not test documents, and certainly we do not test users; we ask organizations and the people whom the organizations serve to help us come up with a solution that they will use in whatever way is appropriate for them. Conversation with users at this early stage of design is of the utmost importance in the collaboration process.

We do not see people as users, audiences, clients, receivers, or targets; these definitions have come from fields such as human–computer interaction or "professional communication" (marketing, public relations, and so on), and they predefine the relationship that exists between people and information. We prefer to ask people what kind of relationship they have or want to have with the information rather than anticipate their relationship to the information.

This stage can involve some numerical work; for instance, figure 3.4 is a data set from a project concerned with designing a telephone bill.

The project was undertaken in 1987–1988, just at the time when high-speed duplex-printing laser printers were being introduced into billing production environments. This particular bill was developed to cope with what was then, and remains, a problem within the telecommunications industry: that people have great difficulty in making sense of the information on their phone bill. Illustrated is one particular data set, which shows the progressive improvement in the bill as it went through five rounds of conversation and refinement.

We asked people to carry out tasks on the bill, such as "Show me what you would do if you had an enquiry about your account" and talked to them about their difficulties. The bill was refined after each round. It took five weeks, one week per round, to complete this stage of the project. All the prototypes used for the conversations were prepared on laser printers. Had we attempted this type of prototyping and iteration with earlier print technology, it would have taken at least twice as long.

This points to one of the significant changes in information design practice since the advent of digital technology and software. In the late 1960s I had incorporated prototyping and iterative design with conversation and refinement as part of information design education and practice (Sless, 1979). But the process was slow and laborious. With the advent of desktop computing and desktop publishing that changed. The change coincided with the establishment in 1985 of the unit that became our Institute, and we were able to incorporate such activity as a matter of routine, refining our methodologies progressively over 15 years and some 200 projects.

	What would you do if you had an enquiry about your account?					Responses to questions about meaning of major bill content	
	Enquiries number at top	Enquiries number at bottom	Ring Telecom	Other	Ring wrong number	Understood all	Understood with exceptions
Round 1	63.2	—	31.6	0	5.3	68.42	31.6
Round 2	65	—	20	15	0	60	40
Round 3	60.5	18.4	7.9	10.5	2.6	76.3	23.7
Round 4	45	45	5	5	0	100	0
Round 5	80	10	5	5	0	100	0

Numbers shown are percentage of respondents

Figure 3.4. Data from the conversation and refinement stage of a project to redesign a telephone bill (Sless, 1992).

Specification and production

There are some interesting differences in the ways we do specifications now to the ways in which we did specifications in the past. Today we increasingly specify the rules for an entire system rather than the appearance of a single object, such as a manual. This kind of design requires a different way of thinking than traditional design where you can focus on the look of a particular thing within a particular page layout and book structure.

Within contemporary information design, there has been a shift away from the Bauhaus tradition of creating and mass-producing copies of a single object, toward developing entire information systems by generating rules enabling us to uniquely customize information for particular individuals (Sless, 1999).

Figure 3.5 is an extract from a single page of a 30-page document, which specifies the elaborate system of rules for building insurance notices.

Here again we see one of the significant contributions of software to information design. However, this too has antecedents. Consider the way in which kerning pairs are incorporated into desktop publishing systems. The computations involved are invisible to the ordinary reader, but are specified and customized for specific letter combinations in particular fonts. Figure 3.5 is an extension of this type of logic, but at a macro level. On top of the unique kerning pair combination for each word, we now have unique combinations at the sentence, paragraph, heading, and document level, with specific unique features for individual readers.

Monitoring

Once the re-designed information is released into the public realm, it must be appropriately monitored. Continuous monitoring is important, to keep abreast of changes in organizations and society. In our experience, few organizations monitor effectively. Many organizations carry out customer satisfaction surveys, which in our view are a waste of time and money; these surveys can collect customer opinion but cannot collect evidence upon which designers can act. Effective monitoring brings the design process full circle; as new problems arise and are identified, the re-designing begins again at the scoping stage.

Tasks and skills

It is possible to look at the stages of collaborative information design from the point of view of the tasks that are completed at each stage and the skills that are needed to complete those tasks. This view of the stages shows one of the most important ways in which this type of information

2.1 Contributions: one

The Contributions page gives Members a total of each type of contribution paid into their superannuation during the year.

In Annual reports, this page also shows Members what they will receive if they leave the plan.

Contributions appear on page 2 of an Annual report and page 3 of an Exit report.

This page shows how to assemble the exit portion of an Annual report. Exit reports do not have this.

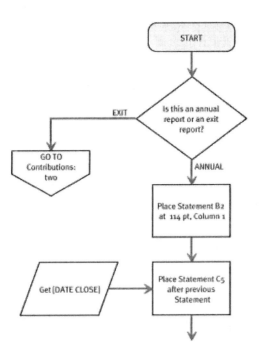

Figure 3.5. Part of a specification for an insurance notices system.

design is collaborative. We put together teams of people with a variety of different skills, and ask them to work on problems with each other (Figure 3.6).

I have observed in the United States and to a lesser extent in Europe that there is often a demarcation between different professional groups involved in information design; technical writers, usability specialists, anthropologists, graphic designers, and software engineers seem to demarcate between each other's professional competencies and tasks.

By contrast, I have been able to break down these demarcations in our information design teams. It is common in our projects for writers and graphic designers to participate in conversations with people using their designs. Equally, those with primary skills in usability or anthropology would be expected to write and design material at appropriate stages. Thus collaboration extends deeply into the design process itself, not just into the collaboration with those who use the designs we create.

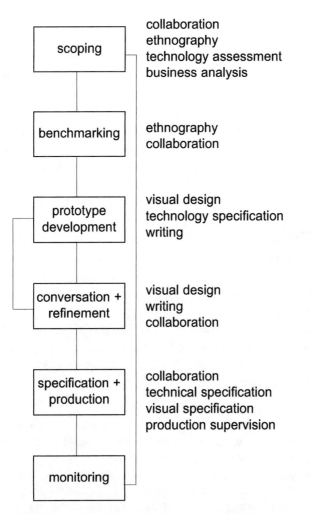

Figure 3.6. Skills involved in each stage of an information design project.

Effort

Figure 3.7 provides an insight into the kind of effort that goes into this work—the amount of time the various tasks take. These data have been taken from a large number of projects for which we have accurate figures.

Two items in particular may surprise the reader. First, notice that only 4% of effort is spent on prototype development. In traditional design practice teaching, prototype development is probably the major element taught, and little is taught about the other aspects. Second, the percipient reader will see that the percentages on the right add up to only 50%. The remaining 50% of the time is spent on politics.

Superficially one can look at the diagram of the information design process and imagine the design process to be rational, even scientific. But it is in fact profoundly political; people's interests and power relationships are an intrinsic part of any information design project. Often, I hear designers say at the end of a project, "It would have been a great project but for the politics." My view is that the politics is not an excuse for failure. Politics are an *intrinsic* part of the process, not an external variable. Dealing with the politics involves a great range of issues. Because politics takes up such a large part of the process, I spend the rest of this chapter exploring part of what that means both in principle and practice. In the remainder of this chapter I deal with two political issues: representing unrepresented constituencies, and shifting problem boundaries. Note that these two issues impinge on, and overlap, each other.

The politics of representing unrepresented constituencies

The incomplete complete solution

The Canary Wharf tube station in London is a triumph of architectural design. It is, if you like, a complete design. It has brought a complete solution into existence. It has reached the edge—its boundary—and within that, it provides a complete solution to the design of a tube station.

One of its dominant features is the brushed stainless steel lining of the entrance hallway, into which are set brushed stainless steel doors with elegant brushed stainless steel handles which invite you to pull them to open. Awkwardly and crudely sticky-taped above one of the elegant brushed stainless steel handles is a sign that reads "Push to Open."

There are other signs stuck on that brushed stainless steel door, and on many others, because, although it is so beautifully and harmoniously designed, the design does not tell tube travelers what it is for.

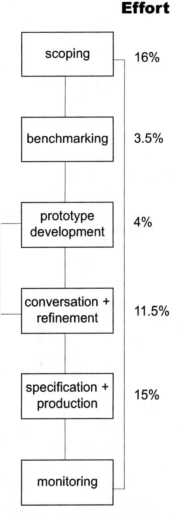

Effort

Figure 3.7. Percentage of effort involved in each stage of an
information design project.

It is clear that the architects' problem boundary was not quite where
it should have been for a complete design; it did not extend as far as the
people who would go through that door. One could argue that during
the scoping stage the full scope of the project was too narrowly defined,
necessitating a later prosthetic fix with sticky-tape.

But such extending of problem boundaries is a profoundly political
process of representation. Clearly, the people likely to use the door were
neither considered nor consulted.

To consult people who have previously never been consulted is to engage in politics. When you import into the decision-making process their opinions and suggestions about the things they have to live with and put up with, you are introducing a formerly unrepresented constituency to that process. This is not an act of usability testing or research, it is a political act. In effect, you are telling those who have exercised power in this area to now give some of that power to someone else. But power is not something that is relinquished easily; total control is not something that people like to abandon.

The professionals

Everyone has views about other people and what they are like. In many situations these views are political—an expression of power relationships, self-interest, and a claim to speak on behalf of others as their representative. In one project, we had to get agreement on benchmark issues from doctors, pharmacists, nurses, other health professionals, and consumer groups. These disparate groups were at first kept separate to avoid fruitless confrontation, because each group had a strongly held professional view about the other groups: doctors had opinions about what doctors knew and could do and what pharmacists didn't know and couldn't do, and so on. Each group was given a list of possible benchmark criteria that we might apply; and strangely, each group's benchmark criteria were almost entirely the same. There were hardly any differences when it came to deciding what they wanted to achieve, despite their professional differences (Penman, Sless, & Wiseman, 1996).

Sex rules okay

In another project, we helped the Australian National University (ANU) to develop sexual harassment information. The ANU had developed comprehensive procedures for dealing with sexual harassment, but they were finding enormous difficulty in making this information not only available to students but also usable by students in an appropriate way.

As you are probably aware, most of the emphasis in such activities traditionally relates to the harassed person. We took the view that everyone involved needed to be informed; thus one of our objectives in this was as much to address people who are at risk of harassing as much as those who feel harassed. This is an example of shifting the problem boundary.

At the beginning, there was much agreement among the various groups about what everyone should know about, and what the procedures were. By the time we got to the benchmarking phase, we knew a lot about what each group wanted and we quickly reached agreement on a set of benchmark requirements for two documents. The first document was to be a brochure handed out during orientation week as part of an information kit for new students. The second was a comprehensive guide to the University's sexual harassment procedures.

Everyone agreed that the first document, the brochure, should enable students to:

- know that the University community has clear and strong rules about sexual behavior,
- discriminate between acceptable sexual behavior and behavior that constitutes sexual harassment,
- be able to seek help and advice or make a complaint about sexual harassment,
- know that some types of sexual behavior are criminal and will be dealt with by police,
- know the potential consequences of sexually harassing someone,
- know their rights and obligations.

Then we developed a prototype, designed to satisfy the list. Figure 3.8 is part of the brochure, which boldly proclaimed on the front cover *SEX RULES at UNI*.

And then the politics erupted. Those who were deeply concerned about issues of sexuality and power told us, with much vitriol, that the title *Sex Rules at Uni* was misleading and ambiguous. Claiming to represent students, they went on to say that the brochure was:

Offensive to students. Unlikely to be taken seriously. The text in general is much too simple, talks down to students, could use more complex language. It sounds authoritarian, paternalistic, inappropriate language

A letter was sent to the Vice Chancellor demanding that the project be stopped, and a committee be set up to do the work "properly."

This reaction was not entirely unexpected. But when we took the brochure to students during the conversation and refinement stage, we found a very different reaction. Unsolicited, in one-on-one conversations, a number of new students told us that they were greatly reassured by the brochure, because it made them less anxious about dealing with this type of problem. When we specifically asked whether they found the content "offensive," "authoritarian," or "paternalistic," all said no. Indeed some of them looked at us as if to say "what a silly question."

Part of our remit was to try to reach students who were at risk of being harassers. Previous brochures, which had sexual harassment in

Figure 3.8. Inside a brochure for students on sexual harassment.

the title, did not interest these students. Hence the ambiguity in the brochure's title. Observation of students going through their information kit found that a number of students did venture beyond the cover of the brochure, and read the inside. However, here, too, worrying politics intruded. Because the brochure spoke even-handedly to all students, the individuals and groups who saw themselves as representing the sexually harassed saw the brochure as inherently biased toward harassers, who in their view had no rights whatsoever.

I am happy to report that, after much negotiation, refinement, and the presentation of the results of our conversations with students, the final brochure was widely accepted and is now used by both students and staff.

Letting voices speak

In this very different type of project, we were concerned with a useful but dangerous medicine, used to treat HIV-positive people. It is part of a cocktail of medicines that keep the immunity system from collapsing. However, it is a drug that is still being developed, and moreover there are known serious side effects, so how the medicine is taken is extremely important. Appropriately taken, it holds AIDS at bay; inappropriately taken, it can kill.

We found that the voice of the person with HIV was most important in this area. However, the official regulations regarding patient information about medicines says that patient information must be written to a certain formal structure and has to follow the Product Information document, which is the document the pharmaceutical company submits to the regulator. The product information does not contain information about the patient, so the consumer information does not contain such information either; it only contains information about administering the drug. We had to find other vehicles for getting the patient's voice in. In this document (figure 3.9) you can see how we have brought in the voice of the person taking the medicine (Mackenzie-Taylor, 1997).

There is no way of letting all the voices speak until everyone concerned is included in a collaborative design process. Negotiating with groups of people with set opinions is an intrinsic part of the information design process, indeed any design process, because it is very much a part of reality. These examples give you an idea of where some of the politics lie.

The politics of boundary shifting

The case histories discussed earlier deal with the issue of political representation. However, as the Canary Wharf and ANU projects

MAKE SURE that you get the most out of indinavir
sulfate (CRIXIVAN).

1   Take it every 8 hours, around the clock every
    day, at about the same times each day.
    Don't skip doses or take double doses.

> I never forget to take my drug cocktail. I
> know it's important to keep a constant
> level of indinavir in my body, making it
> harder for HIV to become resistant.

2   Ideally, do not eat for two hours before or one
    hour after taking your dose. If you need to eat
    during this 'three hour dose window', make sure
    it's only small amounts of light food.   *see pages 23–25*

> I used to think you had to fast for 3 hours
> around taking indinavir. It's a relief to
> know it will still work even if I eat small
> amounts of light food around dose times.

Figure 3.9. Part of 'life, food and Indinivar.'

showed, the issues of representation can lead to another political issue: changes in the nature of the problem boundary. Following are examples of boundary shifting in information design.

Redesigning a hospital signposting system

A large urban public hospital called in an information designer to help redesign the hospital's signposting system. The problem they wanted solved was that of people getting lost and confused when visiting the hospital campus. At the end of the first part of the scoping stage the information designer had discovered that the hospital, which had first opened in the mid-19th century, was made up of a large number of buildings in different architectural styles—gothic revival, utilitarian, modernist, post modern, and an assortment of "temporary" prefabricated buildings from various eras—all squeezed into an increasingly crowded inner urban site with no car parking for visitors. Various generations of attempts at signposting coexisted, some out of date, and many placed in inappropriate positions for hospital visitors. Observations showed that most visitors went up to the majestic vaulted original front entrance—now used exclusively for hospital administration—and tried to find someone to direct them. There were no signposts on this building because it was a listed historical building and such appendages were not allowed. The most frequently asked person was the doorman, whose main job seemed to be to open the majestic doors for consultants. If he was unavailable, visitors would poke their heads around office doors, looking for signs of life.

Moreover, on further investigation, the information designer discovered that the wayfinding problem did not begin at the majestic entrance. Most first-time visitors were either referred to the hospital by a GP or were coming to visit patients, and they had to find their way there by public transport, or struggle to find a parking place and then walk. Visitors' sense of the problem extended a long way outside the hospital boundary. Once they were at the hospital they asked "the nice man in the uniform" for help if they could find him. They went on to say that he gave them very clear directions. But, as one visitor said: "I came across some signposts that said something different. I didn't know what to do, and then I forgot all the instructions he'd given me anyway."

The solution proposed by the information designer was not to re-signpost the hospital—a very expensive option that could only ever work partially on such a complex site. Rather he proposed that the hospital:

- Provide all local GPs with a simple leaflet showing patients how to get to the hospital by car or public transport.
- Take down all the old signs.
- Redefine the doorman's job.
- Provide the doorman with carefully developed maps to all public locations in the hospital to help visitors remember his instructions.
- Design a new map when any change at the hospital affected visitors.
- Let the consultants open the door themselves.

Redesigning a medicine information leaflet

A pharmaceutical manufacturer had developed a medicine and was concerned about a possible side effect if the medicine was not adminis-tered correctly. The medicine—a tablet—could cause upper esophageal burning if it lodged in the esophagus for any time before entering the stomach. They asked an information designer to redesign the medicine information leaflet, instructing patients not to sit or lie down when they took the medicine, and to walk about after taking it. However, during the scoping investigation, the information designer discovered that a num-ber of people taking this medicine had limited mobility and were likely to be unable to "walk about" after taking it. The information designer suggested to the pharmaceutical company that, rather than redesign the leaflet, a more appropriate solution would be to redesign the medicine.

Recording health information in Indian villages

At the invitation of the Indian government, Apple Computers investigated the redesign of the Newton interface (an early handheld computer) for use by health field workers in rural Indian villages. But in the scoping stage it was realized that the technology was totally unsuitable for the cultural context and what was needed was a better paper system.

Some lessons

Four important lessons can be learned from these case histories.

First, the "solution" as originally conceived—a signposting system, a medicine leaflet, a computer—was not the solution that emerged during the scoping stage.

Second, the "problem" as originally stated—people getting lost, not following instructions, having difficulty processing data—was not quite the problem that emerged during the scoping stage.

Third, all the solutions involved a shifting of the problem boundary. In the hospital case the shift was partially physical in that the problem extended away from the majestic hospital doors into doctors' offices and people's homes; in the medicine case, the problem was not in the instructions but in the medicine itself; and in the third case, the problem was inappropriate technology.

Finally, this type of boundary shifting in design is quite common. But it is not simply a technical issue. Such shifts are profoundly political and involve changes in values, implying changes in responsibility, control, ethics, and economics. This can be both rewarding and challenging for information designers. It is at this point that we can see something special about the nature of design problem solving in general, as well as in the specific area of information design. Design is an activity which like much in political life is non-predictable.

It is important to bear in mind that the notion of something being non-predictable is quite different to the scientific idea of something being unpredictable. The representation of the design process as a linear sequence of stages implies, as Simon (1969) suggested, that design is a "science of the artificial." But, in practice, the process is not linear, nor are the outcomes necessarily determined by the starting conditions. In practice, design can be a highly volatile non-predictable activity. This has given rise to some important contributions to design theory, in particular the idea that the class of problems that design tries to solve are "wicked problems" (Buchanan, 1992).

Collaboration, in summary

In the face of all of this, there is nothing remarkable about collaboration. It is the obvious, the only, way to design anything. How could one possibly do it any differently? Why should one want to do it any differently? As designers, we introduce something into the world; therefore we have a social and moral obligation to involve the people who are going to have to put up with it. It is an act of courtesy; it is an act of politeness. I sometimes feel this makes me a professor of the bleeding obvious.

So why do we have to formalize, through all these diagrams, what is a simple act of courteous human behavior? We have to accept that the fact that we have to develop these methodologies—and I'm not against them—is a symptom of the fact that we live in a society where our two great institutional systems, capitalism and bureaucracy, actually do not care about people. I am not suggesting that the individuals in these organizations do not care about people—that is manifestly not the case. What is the case is that the primary purpose of capitalism and bureaucracy has nothing to do with being nice to people. Capitalism is concerned with making money for shareholders, whereas bureaucracy is concerned with control. Whether they are concerned with being nice to people at the same time is another matter entirely. Capitalism and bureaucracies happen to care now, at this historical moment, because at present people have some economic and political power.

No doubt, if capitalism discovers a way of making more money without being nice to people, you can be quite sure that almost every usability lab and user-friendly design in the country will disappear. We must seize the moment, we must develop these things to a point where we cannot go back, we cannot retreat. We need to see what we do clearly, within an historical context. Because it might not last.

Notes

CRIA was established in 1985 as a not-for-profit research organization, set up to help people communicate with each other. CRIA's work is mainly with the corporate sector and large commercial and financial organizations, whose communication practices, both internally and with their publics, are performed badly—they would like to do it better, usually for economic reasons. CRIA's concern, though, is very much with the people who work in those organizations and the people they communicate with.

Acknowledgments

This chapter is about collaboration. When you work in a collaborative environment there are great many people involved; my capacity to write a chapter like this has been made possible only with help and support of others. In particular, I'd like to thank: Dr. Ruth Shrensky of the University of Canberra, who works collaboratively with me in all my undertakings; Professor Stephen Scrivener of Coventry University, and the Conference organizers; Coventry University for providing me with an enormous range of opportunities to share some of the experiences I've had in this field; and my colleagues at the Communication Research Institute of Australia (CRIA). Of course, any errors are my own work.

References

Buchanan, R. (1992). Wicked problems in design thinking. *Design Issues,* *8*(2), 5–22.

MacKenzie-Taylor, M. (1997). Designing for understanding within a context of rapidly changing information. *Vision Plus Review,* 21 E/D. Medieninhaber: International Institute of Information Design.

Marshack, A. (1972). *The roots of civilization.* London: Weidenfeld and Nicolson.

Penman, R., Sless, D., & Wiseman, R. (1996). Best practice in accessible documents in the private sector. In *Putting it plainly: Current developments and needs in Plain English and accessible reading materials.* Canberra: Australian Language and Literacy Council.

Simon, H. (1969). *The sciences of the artificial.* Cambridge. MA: MIT Press.

Sless, D. (1979). Image design and modification: an experimental project in transforming. *Information Design Journal,* *1*(2), 74–80.

Sless, D. (1992). The Telecom bill: Redesigning a computer generated report. In D. Sless & R. Penman (Eds.). *Designing Information For People* (pp. 77-98). Canberra: Communication Research Press.

Sless, D. (1999). The mass production of unique letters. In F. Bargiela–Chiappini & C. Nickerson (Eds.). *Writing business: Genres, media and discourses* (pp. 85-99) Harlow: Longman.

Sless, D. (2001). www.communication.org.au/html/paper_26.html.

4

The Five Dimensions of Usability

Whitney Quesenbery
Cognetics Corporation

Why talk about usability in a book about information design? Although there is not yet a consensus on a single definition of information design, one that I like is making "information accessible and usable to people" (Sless, 1992, p. 1). It is not enough to design well: We must also achieve information design that is usable.

We therefore need to establish what "usable" means, so in this chapter I present five dimensions of usability, which can be used in several ways:

- As a model, they provide a way to understand what kind of usability is needed in different contexts;
- As a tool, they help guide the design process, suggesting both a general approach and specific choices;
- For evaluation, they are both useful as a way of understanding why a design is failing, and suggest appropriate techniques to get the design right.

This multifaceted view of usability allows designers of both complex and simpler products to understand user requirements and evaluate the success of the design.

The work of an information designer shares elements with others including user experience designers, information architects, graphic artists, interaction designers, user interface designers, usability engineers, writers, content managers, indexers, and quite a few more.

Each brings specific skills and their own perspective to the table. Rather than sort out the distinctions, I use the term *designer* generically, and the terms *product* or *interface* for the end result of the design.

Definitions of usability

The problem of defining usability is complicated by the fact that we use the word "usability" to describe:

- *a result* – a quality of a product that is usable
- *a process* – a methodology for creating those products
- *techniques* – the specific methods or activities, such as contextual observation and usability testing, used to achieve that result
- *a philosophy* – a belief in designing to meet user needs

Although all of these are important, this chapter focuses on usability as a result.

So what is this quality of "usability"? To say usability is "ease of use" or even "user friendly" make good sound bites, but the phrases are too simple to communicate the complexity of the user experience. A better definition can be found in an ISO standard on usability (ISO 9241, 1998):

> The extent to which a product can be used by specified users to achieve specified goals with effectiveness, efficiency and satisfaction in a specified context of use.

This definition is more complete, but it reflects the emphasis of much of the original usability work on strongly defined tasks and work tools. This heritage has sometimes led people working in new media from the web to wireless to assume that usability is not relevant to their work. This has been especially true for designers of products without a strong task orientation, such as large information systems. In addition, designers of consumer products criticize usability as not accommodating the concept of "fun" as a goal, even though it can be important in the marketplace.

For usability to apply to an experience in which a human can interact effortlessly with a product, its definition must encompass both work metrics and a sense of delight, wonder or exploration.

Five dimensions of usability

To create a new model, I expanded the ISO 9241 characteristics of usability (efficient, effective, and satisfying) to five dimensions: effective, efficient, engaging, error tolerant, and easy to learn (Quesenbery, 2001).

"Satisfying" becomes "engaging" not merely to preserve the alliteration, but to raise the emotional level and create a sense of a dynamic interaction.

These dimensions, the 5 E's, each describe an aspect of the user experience (figure 4.1, next page). Taken together, they are a tool to create a more precise description of both the goals for, and experience of, using a product. This chapter looks at how this broad view of usability contributes to a successful product.

As a prescriptive design tool, the relative importance of each dimension can be specified, allowing the usability goals to be balanced appropriately for each context. Although this flexibility offers no single interpretation of "usability," it offers the opportunity to customize its meaning for each product, based on the needs of the users.

If the unofficial motto of usability is "it depends," this model provides a definition of what "it" depends on. This allows designers to avoid simplistic rules which may or may not be appropriate. For example, the often-quoted metric that "anything on a web site should be reachable in three clicks" may or may not apply to any individual site and its users. It also allows business goals and other constraints to be factored into design decisions, balancing function requirements with usability requirements.

An analysis of the 5 E's can start a human-centered design process. They provide a clear statement of user goals to complement other contextual and task analysis. This chapter describes how to define these goals, but a complete description of user-centered design is beyond its scope. A list of uncited references can be found in Further Reading under "Usability as a Process."

The usefulness of the 5 E's does not end with understanding users. As a descriptive tool, the five dimensions continue to be helpful. They suggest design approaches and can then be used to evaluate why an interface is succeeding or failing.

Effective

The completeness and accuracy with which users achieve their goals.
Effectiveness looks at how well the product helps people reach their goals. A user's goal might be to find all important information on a topic, have fun, or complete a clerical task with minimum effort. Effectiveness is measured by finding out whether the user's goals were met successfully.

The details of these goals can vary dramatically in different contexts:

- Users of a reference web site or corporate intranet used as a document repository might need to find all relevant documents on a topic, rather than simply selecting a few likely candidates.

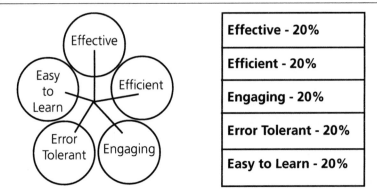

| Effective - 20% |
| Efficient - 20% |
| Engaging - 20% |
| Error Tolerant - 20% |
| Easy to Learn - 20% |

Figure 4.1. The five dimensions describe different aspects of usability. This diagram shows them in balance. However, in most products some dimensions have a higher priority than others. The challenge of usability is to create a design that successfully accommodates all of the dimensions. (This diagram concept was developed in part through private communications with C. Jarrett, August 2001.)

> In this case, the completeness of the search is important, lest critical information be overlooked.
> - A customer service representative may need to complete an order form so that it can be fulfilled immediately. All entries must be correct, all necessary information present, and there must be no ambiguities needing explanation.
> - A help system must be able to provide the correct information for the user to continue working effectively with the minimum interruption of their task.
> - For games, the goal is to have fun. A successful game challenges the player at just the right level.

Effectiveness does not require an annoying or dictatorial interface which forces users down a single path toward the "right answer, " but an understanding of their goals and the information and flexibility needed to reach them.

Efficient

The speed (with accuracy) with which users can complete their tasks.
ISO 9241 defines efficiency as the total resources expended to complete a task effectively. Resources include the number of individual actions a user must take and the time spent on them.

It is important in measuring efficiency that the boundaries of the task are defined appropriately. The user's perception of the complete task, rather than individual functions as they are organized in the product, should be used for this definition. This is especially critical when a task involves multiple functions or when the entire task cannot be completed within the product.

- For an example, let us go back to a corporate user and the goal of finding all relevant documents. If we set the boundaries of the task as determining how much time it takes to enter a search term, execute a single search and retrieve a document, this system might appear very fast. But, if we look at the time the user takes from deciding to search to locating the documents they need, the results might be quite different.
- The design of the navigation and the categorization of the content also has an impact on efficiency. In a work application like the one used by the customer service representative, an interface that makes it easy to move around the functions might be more efficient to use than one with a hierarchical or linear structure, because its use will require less (inefficient) backtracking, or repetitive traversal of a hierarchy.
- In a help system, dead-end topics can be converted into alternate navigation choices with appropriate links, enabling the user to proceed without delay. Examples of this are links to additional information included with a topic, or links to other closely related topics.

User assistance has a specific efficiency requirement. Because contextual help or other integrated user assistance interrupts the user, it must take as little time as possible to guide the user to the answer to a question. Clear language also contributes to efficiency. Any word that forces the user to stop and think adds friction to the interface, whether it is an unfamiliar technical term or a cute label that conveys little meaning.

Often, effectiveness and efficiency are closely entwined. In many cases a marginal increase in efficiency may be less important than completing a task correctly. Understanding the relative value of these two dimensions to users is an important step in design.

Unless the context in which the product is used demands an absolute minimum time for use, a perception of efficiency may be more important than actual timings. If they are able to navigate without backtracking or error, for example, an interface with a deeper hierarchy (i.e., one that requires more clicks) may seem faster to use than one where users become confused or lost.

Engaging

The degree to which the tone and style of the interface makes the product pleasant or satisfying to use.

An engaging product is one that draws the user in, encouraging interaction. It is the most subjective of the five dimensions. Visual design is the most obvious medium through which the product attracts its audience, but the choice of language, the media used, and the style of the interaction all play a part in creating the experience that creates engagement.

- A help system or other information site might show glimpses of other useful topics, promising a continuing interest. One technique for doing this is the use of "related topics" links, signposts that point to new pages. These may take many forms, but all of them rely on the idea of browsing—allowing the user to follow a line of interest from topic to topic.
- The organization of the information and the way that organization is communicated to the user can also affect how engaging a product is. For example, on an intranet, the site map, table of contents, or index must be presented in a way that helps the user understand it. Info-graphics or information visualizations used to communicate a complex concept in a simple form are both examples of engaging interfaces. They may even be fun to use.
- Finally, the initial view—home page, front panel, or main menu—must communicate the contents and scope of the product. This initial impression must carry a lot of weight: The visual style must be appropriate to the business setting, there must be enough information to help the user form a mental model, and the options for interaction must be clear and compelling.

It can be more difficult to see how a customer service application can be engaging, but this dimension is important here too:

- An application that the user is required to use can be pleasant or unpleasant. A product that is attractive, respectful, and helpful is more engaging than one which is ugly, rude, or aggravating.
- A good visual layout makes it easier for the user to work with the customer by providing the information they need in a highly readable form.
- Small rewards built into the application can congratulate the user on reaching a work target.

Error tolerant

How well the design prevents errors, or helps with recovery from those that do occur.
There are three sources of errors:

- There may be defects in the code or unanticipated conditions that interfere with the operation of the program.
- The designers may have misunderstood the user's mental model and created an interface that does not work predictably.
- The user may simply make a mistake.

An error-tolerant system is designed to help the user recover from such problems by providing information, choices of actions to correct the problem, or other solutions. It also accommodates accidents, misinterpretations, or other deviations from a single path to the goal.

Defects are often caused by a failure to consider all the possibilities for user interaction. Strategies for error tolerance include:

- Errors such as missing information or data in an unanticipated format should be treated as part of the normal process and their correction incorporated into the interface rather than handled as an exception in a separate subsystem.
- It should be difficult for a user to take invalid actions. Solutions include providing selection lists of appropriate choices, providing clear examples for data entry, or disabling options when they are not active.

When errors are caused by a mismatch between user expectations and the system's technical model, the strategies used need to restore control to the user as quickly as possible.

- Providing feedback as close as possible to the action provides the user with an opportunity to correct any mistakes immediately, so they can be corrected before their attention has shifted away.
- Making a product easy to learn also increases its error tolerance. Difficult tasks should be designed to include basic instructions, reducing the chance of an error based on confusion.

The final type of error is rarely considered carefully enough: The user might simply make a mistake.

- People can mis-click, selecting a different item than they intended. Poor navigational cues might lead them to choose the "wrong" item. To make these situations error tolerant, all actions should be reversible. This includes the ability to back-track a navigational choice, or to undo an action easily.

- There are also situations in which the choice might be ambigu-ous with more than one "correct" selection. These are not errors in the classic software development sense, but they do result in a mistake from the user's point of view. In this situation, the degree to which the design is error tolerant has a direct impact on its effectiveness.

The language in messages, especially error messages, can make the difference between a product that seems rude and unhelpful and one in which "errors" are transformed into "navigational options." A well-written message will not only help the user recover from a problem, but also help prevent them from repeating the problem in the future.

Needless to say, documentation or user assistance should not pro-duce errors, especially when it is used to help users recover from them. This places special demands on online training or documentation to work flawlessly.

Easy to learn

How well the product supports both initial orientation and deepening under-standing of its capabilities.
A product that is easy to learn allows users to build on their prior knowledge without deliberate effort. Ease of learning is created through some specific techniques, but it is also embodied in a helping attitude throughout the interaction.

Good instructions, prompts, examples, or hints can provide enough information to create an interface in which the user can extend prior experiences into a new context.

As a dimension of usability, ease of learning includes new discovery that goes on for the complete life of a product. For example, as they use an intranet, people expand their scope of work, explore new options, require access to additional functionality, or even change the way they interact with the product completely. Whatever the cause of these changes, the continued usefulness of the product depends on its ability to support the user through them.

One of the criticisms of usability is that it places too much weight on the ease of initial learning, or so-called intuitiveness. Many usability tests observe users' initial encounters with the product as a way of exposing problems that can be easily masked once the interface becomes familiar. Critics complain that this methodology tends to produce products with a low barrier to entry, but which are not powerful enough for sustained use, or become tedious once the user is familiar with them. For example, a screen-full of instructions in an application that were helpful on first use can easily become annoying when it must be scrolled past every single time the page is read. Worse, long-time users of products such as the customer service application become desensitized to messages or prompts, and are more likely to make an error by not reading another instance carefully. An example of this phenomenon is the tendency to click "OK" on every confirmation dialog without reading it.

An easy to learn interface is both consistent and predictable. A consistent interface ensures that terminology does not change, that design elements and controls are placed in familiar locations, and that similar functions behave similarly. Predictability expands this to place information or controls where users expect them to be, and act in ways they expect (Bergman & Haitani, 2000). A predictable interface allows users to transfer prior knowledge from similar products, and to make use of any interaction patterns they may have experienced.

Advanced functions, such as keyboard shortcuts, menu type-ahead, or other ways of compressing a hierarchy can dramatically increase efficiency, but must also be easy to discover and thus take little effort to learn. In some kinds of games, the discovery of these shortcuts is part of the fun, creating a tension between "engaging" and "easy to learn."

The 5 E's are interdependent

All 5 E's must be considered together, and the product must provide for user needs in each dimension.

There are natural relationships that can be exploited. We have already discussed how efficiency and effectiveness work together. Another example is the way that good instructions built into the interface help improve both ease of learning and error tolerance.

There are also some natural tensions that must be balanced for a successful product. When one dimension is dominant, it is easy to ignore others, especially when they introduce competing requirements. The long-running debates between visual designers and usability engineers are partially a result of failure to explore ways to create a user experience which is both engaging and efficient. Each group concentrates on a single dimension rather than looking at the whole user experience.

Using the 5 E's to set design priorities

Understanding the 5 E's is useful as a way of seeing the multifaceted nature of usability, but their real importance comes when they help to set priorities for the product's design. As a tool, they can be used throughout the human-centered design process for both analysis and evaluation.

We can start by examining the five dimensions of needs for the users of a product. Each group of users should be considered separately, and then compared to the others for similarities and differences.

Contrasting priorities between applications

The following two examples show a summary of user needs for two products: an online photography exhibition (figure 4.2) and the museum web site, contrasting the different requirements between them.

Online museum exhibition

To accompany an exhibition, a photography museum created a web site with samples of the images, information about the artist and about the exhibition. The museum wanted to both attract more visitors and to provide a long-term educational site. The primary target users were: tourists looking for exhibitions, people already interested in the artist, and casual visitors linking from the museum site for additional information.

Effective	The content of the site must be effective in communicating the exhibition material. Questions about the artist and the museum exhibit must be easily answered.
Efficient	This is not a primary concern. People browsing photographs are less interested in how quickly they can move around the site than in the richness of the experience. However, the size of the images might be a problem and long downloads needed to be avoided.
Engaging	The site needs to be engaging in several ways: to encourage those unfamiliar with the artist to stay and explore; to provide new and interesting information for researchers; and to create a compelling experience in its own right as an exhibition.
Error Tolerant	Any content errors are unacceptable. In addition, the rich media used on the site created several opportunities for problems.
Easy to Learn	One of the goals of the site is to encourage discovery. It must therefore invite exploration.

General museum web site

The museum also has a general web site, with information about their exhibits, educational programs, awards, and other activities. The user groups for the larger site are more diverse than those for the exhibit,

Figure 4.2. The photographer Weegee was a character in his own right. The web site for this online exhibition opens with an image of him, with his camera, looking directly at the visitor. www.icp.org/weegee. Source: © International Center of Photography.

including: the same tourists, people shopping in the museum store, job seekers and art world colleagues keeping up with the institution. All of these users are seeking information about the museum, though the details might be different. Their usability requirements are all similar:

Effective   The site must include content that answers users' questions in an easy-to-find location.

Efficient   Attention spans are relatively short. The site structure must be straightforward and direct to minimize navigation time. Writing should be concise and easily scanned.

Engaging   For the users, the site provides their first impression of what the museum is like. The degree to which the site can delight the visitor (and by extension convince them to visit the actual museum) is a measure of success.

Error Tolerant   Errors are not acceptable in any form, especially those caused by a failure to meet user expectations.

Easy to Learn   Users do not expect to have to learn to use a site. This site must allow for "zero-trial learning"—the ability to just walk up and use a product successfully the first time.

The diagrams in figures 4.3 and 4.4 visually present the dimensions in proportion and shows how the usability requirements differ for these sites.

Varying requirements within a product

Sometimes, different users may have conflicting needs that must be resolved in the design. Individuals may also have varying needs, depending on the function. And needs can change over time or as the context changes. In analyzing a more complex product, you need to look for unexpected overlaps or differences. Requirements are not always straightforward to analyze.

Payroll application
When users of a payroll application were interviewed, their priorities varied depending on the function being described:

- For routine functions, such as entering weekly time sheets, they wanted a very efficient interface. This work was repetitive and had to be performed every week.
- For difficult or infrequently used functions, such as issuing a replacement check, they wanted the function to be easy to learn: a guided experience with all options explained carefully to be sure they were doing the work correctly.

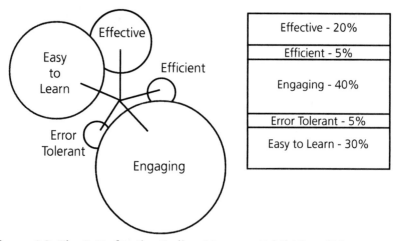

Figure 4.3. The 5 E's for the Online Museum Exhibition: When one dimension is significantly more important, it is easy to lose sight of the others. In this design, efficiency and error tolerance need special attention to ensure that failures in these dimensions do not undercut the overall success of the site.

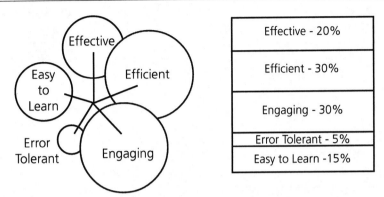

| Effective - 20% |
| Efficient - 30% |
| Engaging - 30% |
| Error Tolerant - 5% |
| Easy to Learn -15% |

Figure 4.4. The 5E's for the General Museum Site. In contrast to the exhibition, the museum site has more balanced usability require-ments. Although error tolerance is a low user priority, this often means that users simply expect it to be there, not that it can be ignored in the design and development of the site.

- Throughout the interface, they wanted error tolerance. They wanted to rely on the software to check their work for prob-lems, and to help them enforce legal and business rules.

Fortunately, the pattern of which functions were considered "rou-tine" and which ones "difficult" was generally consistent among differ-ent user groups—and corresponded to the impressions of the technical support staff from the phone calls they received. In this case, the user requirements, user goals, and business goals were in harmony: Everyone wanted an accurate payroll which could be created and maintained with as little overall effort as possible.

Identifying design tactics

After identifying design goals, the next use of the usability dimensions is to suggest design approaches or tactics. The goal at this stage is to identify the design elements suggested by the usability needs, and see how they fit together.

Table 4.1 shows some typical needs and tactics for each of the dimensions. It contains some of the heuristics I use as the starting point for a more detailed analysis.

Notice that these tactics are not specific design implementations. Instead, they are general principles that act as a starting point for a design. Their value is ensuring that the design decisions are based on user needs, not which technique is the most fun for the product team. There may be many ways to implement each requirement. The goal of the design work is to find a way to fit all of them together into a coherent whole.

An example of how a user need can suggest a design tactic, which in turn becomes a feature of a product, can be found below.

American Journey

American Journey (Primary Source Media, 1994) is a series of collections of historical images and documents for high school students. It is used as part of a curriculum on how to conduct research. The design team identified a user group of "Reluctant Researchers"—students who were typically not interested in school subjects and needed to be drawn into the material.

Our solution was to look for a way to engage them, and we decided to use the images, which were very attractive. We knew that the solution needed to be easy to learn, so we created an album of the best pictures, which could be easily browsed (figures 4.5 and 4.6).

To integrate the Picture Album into the design, we used it as one of the entry points. Each image had a link to its main entry in the database. From this screen, the student could continue browsing to find other, related, images and documents—beginning the exploratory research process.

Although it was initially designed to solve the usability requirements of one user group, the solution became a feature of the product, absorbed seamlessly into the overall design.

Creating usability goals

The 5 E's continue their usefulness as part of creating usability goals for a product. By connecting the original user requirements, they ensure that the usability goals express user needs well.

A usability goal is a design objective that is unambiguous and measurable. Usability goals guide the design process by establishing the most important values and the objectives a product must meet. It is important that they be accepted by the entire development team; otherwise they have simply deferred the inevitable conflicts, possibly to the point where they cannot be resolved successfully.

A well-written goal has four components:

> *User Definition:* Which users does this goal apply to?
> *Task:* What should they be able to do?
> *Context:* Under what conditions does the goal apply?
> *Criteria:* How will the success of this goal be measured?

Table 4.1. Design approaches to meet key usability requirements

Dimension	Key Needs	Design Tactics
Effective	Accuracy	Consider how many places in the interface are opportunities for error, and protect against them. Look for opportunities to provide feedback and confirmations.
Efficient	Operational speed	Place only the most important information in front of the user. Work on navigation that moves as directly as possible through a task. Be sure the interaction style minimizes the actions required.
Engaging	Draw users in	Consider what aspects of the product are most attractive and incorporate them into the design.
Easy to Learn	Just-in-time instruction	Create step-by-step interfaces to help users navigate through complex tasks. Look for opportunities to provide small chunks of training.
Error Tolerant	Validation	Look for places where selection can replace data entry. Look for places where calculators can support data entry. Make error messages include opportunities to correct problems.

It is important to create specific metrics rather than using general criteria. For example, if users say "it has to be quick," we know that efficiency is important. But do they mean they need to complete the task in seconds or minutes?

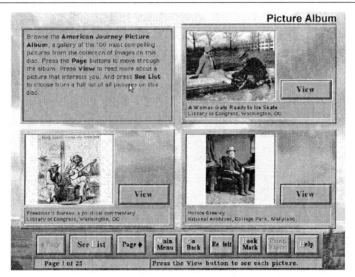

Figure 4.5. The Picture Album displays thumbnail images. Page buttons allow easy browsing through the album. Links lead to a larger image and full descriptions. These pages include links to related documents, enticing readers into the main content of the disc. Source: © 1994, Gale Group Inc.

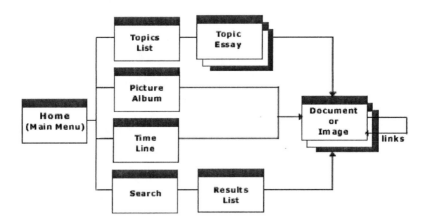

Figure 4.6. The Picture Album provided another entry point to the content of the American Journey database of primary source historical images and documents.

Conference registration site

Users of a conference registration system want their registrations to be accurate and complete. In other words, they have high needs for effectiveness and error tolerance. The conference organizers want to be sure that user satisfaction with the new system is high. The usability goals for the site are derived from both the user requirements and business goals:

Effective
User Requirement: I want you to register me without error.
Business Goal: Reduce costs in processing registrations
Usability Goal: Fewer than 5% of the registrations will have errors, omissions, or inconsistencies requiring a follow-up contact by the staff.

Efficient
User Requirement: I want to get this over with quickly.
Business Goal: Reduce costs in processing registrations, and staff time to manage incomplete forms.
Usability Goal: The user will be able to successfully complete the registration in under 5 minutes

Engaging
User Requirement: I want to feel that the conference organizers care about me.
Business Goal: Convert to a completely online system without customer objections.
Usability Goal: In a follow-up survey, at least 80% of registrants will express comfort with using the online system instead of a paper form or phone registration.

Error tolerant
User Requirement: I want to be sure I do this correctly. Mistakes might cost me money.
Business Goal: Reduce the number of registrations that have to be updated manually to correct mistakes.
Usability Goal: The system will validate all housing, meal and tutorial choices and allow the user to confirm pricing for these options before completing the registration.

Easy to learn
User Requirement: What's to learn about registering for a conference?
Business Goal: Ensure that savings from the new online system are not eliminated by costly technical support.
Usability Goal: Users will be able to successfully complete the registration at the first attempt.

Usability goals are used as part of the envisioning work for a product, to help make decisions during the design process, and at the end when the design is evaluated.

Table 4.2. Evaluation techniques for the dimensions of usability

Dimension	Type of Evaluation	Requirements
Efficient	Timed work on realistic tasks, counting clicks/keystrokes or elapsed time	High fidelity prototype or working product with realistic data available
Effective	Analysis of navigational paths to see how often users made good choices.	Low or high fidelity prototype, with all navigational options enabled
	Tasks which can be evaluated for how accurately they were completed, and how often they produced errors or other problems	Prototype which includes adequate representation of all elements in the tasks
Engaging	User satisfaction surveys or qualitative interviews to gauge user acceptance and attitudes towards the software These evaluations may be done over time, or at key points in use (such as initial experience, short usage, longer usage)	Sufficiently high-fidelity prototype to represent final product Or Working product
Error Tolerant	Include tasks with ambiguities or other potential problems in test scenarios	Working code, or prototype with good representation of error-handling routines
Easy to Learn	Control how much instruction is given to test participants Include functions or tasks with varying degrees of difficulty or familiarization	Any medium or high fidelity prototype, including any on-screen instructions

Using the 5 E's to validate a design

Once the design is complete, the 5 E's continue to be useful as the interface is evaluated. Their characterization of the usability needs for a product helps determine the specific kinds of usability evaluation necessary to validate a design (table 4.2; see left).

Relationship to usability heuristics

One of the common methods of usability evaluation is a heuristic review (Molich & Nielsen, 1990). These reviews use a set of design principles to identify problems that users might encounter in using the product. A typical list of heuristics (Cognetics, 2000) includes:

- Matches users' mental model
- Speaks in the user's language
- Appropriate visual design and layout
- Consistency
- Visibility
- Support for user actions
- Prevents errors
- Includes shortcuts
- Supports discovery and learning
- Provides user assistance
- Supports standards

If heuristics are descriptive, the dimensions of usability are goal-oriented. Together they provide insights into both *what* the problems are, and *why* they are a problem. Table 4.3 shows some of the relationships between the heuristics and the 5 E's.

No matter what evaluation technique was used, a careful analysis of usability flaws against the five dimensions increases the value of the evaluation. The five dimensions of usability offer a direct link to the original usability goals and therefore to any underlying problems in the design approach which are causing the usability problems.

Conclusion

Information is designed for people to read and use. Usability is concerned with how well they can do so. It is the measure of how successful a design is in "making information accessible and usable."

In this chapter, we have discussed the 5 E's:

- **Effective**. The completeness and accuracy with which users achieve their goals
- **Efficient**. How directly and quickly those goals can be met, or the speed (with accuracy) with which users can complete their tasks
- **Engaging**. The degree to which the tone and style of the interface makes the product pleasant, satisfying or enticing to use
- **Error tolerant**. How well the design prevents errors, or helps with recovery from those that do occur
- **Easy to learn**. How well the product supports both initial orientation and deepening understanding of its capabilities

These five dimensions of usability offer information designers a way to define user requirements in a way that can help analyze, design, and evaluate an interface. This model provides a way of understanding the relationship between the content, and its presentation and use, that can guide the creation of the visual presentation, information design, and navigation structure as a unified product that meets user needs.

References

American Journey (1994). Woodbridge, CT: Primary Source Media

Bergman, E., & Haitani, R. (2000). Designing the palm pilot. In E. Bergman (Ed). *Information Appliances and Beyond* (pp. 81-102). San Francisco: Morgan Kaufmann

Cognetics Corporation (2001). Cognetics design guidelines. Retrieved August 1, 2001 from www.cognetics.com/services/heuristic_guidelines.html

ISO 13407:1999 (1999) *Human-Centred Design Processes For Interactive Systems*

ISO 9241-11:1998 (1998). *Ergonomic requirements for office work with visual display terminals – Part 11: Guidance On Usability*

Molich, R., & Nielsen, J. (1990). Heuristic evaluation of user interfaces. Presented at ACM SIGCHI Conference on Computer Human Interfaces.

Quesenbery, W. (2001). What does usability mean: looking beyond 'ease of use. Presented at Society for Technical Communications Annual Conference. Retrieved August 1, 2001 from www.cognetics.com/presentations/whitney/more-than-ease-of-use.html

Sless, D. (1992). What is information design? In D. Sless & R. Penman (Eds.), *Designing information for people* (pp. 1-16). Canberra: Communication Research Press.

Table 4.3. Matching heuristics to the dimensions of usability

If you have problems in these heuristics	Consider whether the design is meeting usability goals in this area
Consistency, visibility, includes shortcuts, speaks in the user's language	Efficiency Are users working harder (and more slowly) than necessary?
Support for user actions, matches users' mental model, prevents errors	Effective, Error Tolerant Are users encountering problems understanding the interface or working completely and accurately because the product is not in sync with their expectations?
Appropriate visual design and layout	Engaging Is the design inviting and helpful?
Supports discovery and learning, provides user assistance	Easy to Learn, Engaging Are these problems caused because the assistance is not available, because it has not been found, or because it is not actually helpful?
Supports standards	Error Tolerant, Easy to Learn Are users making mistakes because they expect the design to follow a standard? Or because they are expected to already know a standard?

Further reading

Usability as a process

Beyer, H., & Holtzblatt, K (1998). *Contextual design: Defining customer-centered systems*. San Francisco: Morgan-Kaufmann.

Cognetics Corporation (2000). The LUCID framework. Self-published. Retrieved from www.cognetics.com /lucid.

Hackos, J., & Redish, J. (1998). *User and task analysis for user interface design*. New York: Wiley.

Morville, P., & Rosenfeld, L. (1998*). Information architecture for the World Wide Web*. Cambridge: O'Reilly.

Wood, L. (Ed). (1998). *User interface design: Bridging the gap from user requirements to design.* Boca Raton: CRC Press.

Usability as techniques

Dumas, J. & Redish, J. (1999). *A practical guide to usability testing* (2nd ed.). Exeter, UK: Intellect.

Rubin, J. (1994). *Handbook of usability testing.* New York: Wiley.

Nielsen, J. (1993). *Usability engineering.* Boston: Academic Press.

Usability as a philosophy

Krug, S. (2000). *Don't make me think.* Indianapolis: New Riders.

Landauer, T. (1995). *The trouble with computers: Usefulness, usability, and productivity.* Cambridge, MA: The MIT Press.

Norman, D. (1990). *The design of everyday things.* New York: Doubleday.

5

Applying Learning Theory to the Design of Web-Based Instruction

Susan Feinberg
Margaret Murphy
John Duda
Illinois Institute of Technology

To create web-based instruction, current practice for most academic institutions includes using lectures, PowerPoint slides, and demos for "porting" to the internet. An interdisciplinary team at Illinois Institute of Technology (IIT) used a more theoretical approach to create web-based instruction. As part of an interprofessional project, the team researched and applied cognitive load learning theory to the design of a web-based instruction module. Cognitive load learning theory uses information design, the design of external representations, to amplify cognition.

Based on recent findings by cognitive scientists John Sweller, Graham Cooper, Paul Chandler, and others mainly in the field of educational psychology, cognitive load is defined as the amount of "mental energy" required to process a given amount of information. As the amount of information increases, so does the associated cognitive load on our mental resources. When the amount of information and instruction exceeds the capacity and limitations of our mental resources, learning will be inhibited.

The basic premise of cognitive load theory is that the focus of an instructional module must be the instruction itself (i.e., intrinsic cognitive load). Information that is adjunct to the instruction (i.e., extraneous cognitive load) must be designed to reduce cognitive demand. Cognitive load theory provides general design principles to: a) reduce the extraneous cognitive load caused by the format of the instruction, and b) increase the capacity of working memory.

This chapter presents an overview of the model and key principles of cognitive load learning theory; an application of these principles to the design of actual web-based instruction; a description of the design and results of a usability test of the instruction; a redesign of the instruction based on user test results; and, finally, a checklist for incorporating cognitive load learning theory in the design of effective instruction.

Cognitive load learning theory

A brief summary of the cognitive load learning theory developed by Professors John Sweller, Graham Cooper, Paul Chandler, and other researchers at the School of Education Studies, The University of New South Wales, is provided in this section as the theoretical basis for the web-based instruction developed by the IIT team.

Modal model of memory

Cognitive load theory describes an information-processing system that involves a modal model of memory (Cooper, 1998). The modal model of memory distinguishes among three distinct memory types (modes): sensory memory, working memory, and long-term memory. Each mode has its own characteristics and limitations that impact directly on the learning process. The integration of these three modes defines the information processing model of human cognitive architecture (figure 5.1).

Sensory memory
Sensory memory deals with incoming stimuli from our senses, including sights, sounds, smells, tastes, and touches. A separate partition of memory exists for each of these senses. Sensory memories extinguish extremely quickly (about half a second for visual information and 3 seconds for auditory information). In that time, we must identify, classify, and assign meaning to the new information or it will be lost.

Working memory
Working memory (previously named short-term memory) is the part of our mind that enables us to think (both logically and creatively), to solve

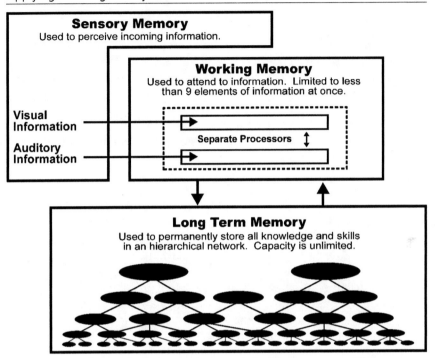

Figure 5.1. Information processing model of human cognitive architecture. Source: Cooper, 1998, p. 3.

problems, and to be expressive. Working memory is directly related to how we direct our attention to "think about something." The greatest limitation to working memory is its capacity to deal with no more than about 7 ± 2 elements of information (digits or subjects) simultaneously (Miller, 1956). Miller's research was the first attempt to quantify the severe limitations that we have on our ability to receive, process, and remember information. This research also recommended avoiding the "information bottleneck" by organizing stimulus input into a sequence of chunks.

Further research was conducted on short-term memory and the processes that transform information from short-term to long-term memory (fixation). The "chunk" capacity was shown to be in the range of 5 to 7, with fixation taking about 5 to 10 seconds per chunk (Simon, 1974).

Total learning time was found to be proportional to the number of chunks of information to be assembled. The psychological reality of the chunk has been well demonstrated since these initial studies and provides the basis today for presenting information and understanding both simple tasks and more complex cognitive performance.

The term *working memory* evolved from its predecessor *short-term memory* as a result of research that characterized this brain system as not only a storage facility, but also a manipulation system for the information necessary for complex cognitive tasks (Baddeley, 1992). Moreover, working memory was disproved as a unitary system, and modeled instead as a three-part system (figure 5.2). This system includes a central executive system that acts as the attention-controlling system, and two slave systems: the visuospatial sketch pad that manipulates visual images, and the phonological loop that stores and rehearses speech-based information.

Long-term memory
Long-term memory refers to the immense body of knowledge and skills that we hold in a more-or-less permanently accessible form. Most everything that we "know" is held in our long-term memory in the form of schema (figure 5.1). Activation of long-term memory occurs as a direct result of our working memory querying long-term memory for specific factual information. Knowledge and skills that are activated regularly, such as walking and talking, may be activated automatically without the need for high levels of conscious attention. The capacity of long-term memory appears to be unlimited.

Primary learning mechanisms: Schema acquisition and automation
A schema is a mental construct that permits problem solvers to catego-rize problems according to solution modes. Knowledge and intellectual skill that is based on knowledge are heavily dependent on schema acquisition. Schemas enable experts to categorize problems and solve them. Novices, who do not possess detailed schemas, are unable to categorize problems. Schemas are therefore the determinants of expert performance (Sweller, 1994).

Along with schema acquisition, automation is also a mechanism for learning. Automation of procedural knowledge deals with the development of skills. Processing of information is either controlled or

Figure 5.2. Baddeley and Hitch working memory model. Source: Baddeley, 1992, p. 557.

automatic. Controlled information is being consciously attended to; no conscious control is involved in automatic processing. The switch from controlled to automatic processing is continuous and slow.

Schemas and automation appear to explain the differences between experts and novices. Together, schema acquisition and automation allow us to bypass or mitigate the restrictions of working memory.

Principles of cognitive load learning theory

Cognitive load theory grew from the study of the cognitive phenomenon that distinguishes between experts and novices and gives rise to speed and accuracy of expert performance. A key premise of cognitive load theory is that technical material must be structured to facilitate schema acquisition.

If schemas determine expertise, good instructional design is that which facilitates schema acquisition. Schemas have been successfully acquired once they are stored in long-term memory for later retrieval and application. They are acquired only after they have been attended to and processed by the cognitive resources available in the working memory (Baddeley, 1992). Cognitive load theory suggests that instructional materials facilitate learning by directing cognitive resources toward activities related to schema acquisition. Therefore, effective instructional material promotes learning by directing cognitive resources toward activities that are relevant to learning rather than to processes that are adjunct to learning.

Intrinsic and extraneous cognitive load
Cognitive load theory distinguishes between two types of instructional information: intrinsic cognitive load and extraneous cognitive load.

Intrinsic cognitive load is related directly to the difficulty of the to-be-learned content (Sweller, Chandler, Tierney, & Cooper, 1990). Intrinsic cognitive load is characterized in terms of "element interactivity" and is not manipulatable (i.e., it cannot be modified by instructional design). Extraneous cognitive load is defined as any cognitive activity engaged in because of the way the task is organized and presented, not because it is essential to attaining relevant goals. Many learning and problem-solving activities impose a heavy extraneous load. By changing the instructional materials presented to students, the level of extraneous cognitive load may be modified, thus facilitating learning. The relationships between intrinsic and extraneous cognitive load, total cognitive load, and mental resources are illustrated in figures 5.3 and 5.4.

Sweller et al. focused on the number of elements (learning items in their simplest form) that the learner must simultaneously attend to as a

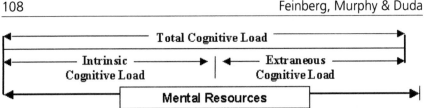

Figure 5.3. Cognitive load Model I: Mental resources exceed total cognitive load   (Learners can learn from any type of instruction.) Source: Cooper, 1998, pp. 11-12.

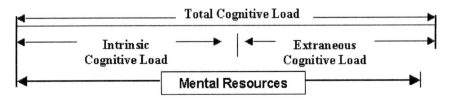

Figure 5.4. Cognitive load Model II: Cognitive load exceeds mental resources . (Learning is facilitated by reducing extraneous cognitive load.) Source: Cooper, 1998, pp. 11-12.

key to understanding cognitive load. Simple content with relatively few intrinsic interactive elements is not threatened by weak instructional methods. Content that is characterized by high levels of interactivity among its elements (grammar, algebra, geometry) cannot be learned effectively through weak instructional methods. Cognitive load theory is designed to characterize extraneous cognitive load so that designers can modify instruction to minimize this load. Much of the cognitive load theory is targeted toward teaching math and science units using the goal free effect (students draw all possible conclusions from given data), the worked example and problem completion. Three additional techniques include the split-attention effect, the redundancy effect, and the modality effect. These three approaches have direct application in the generation of web-based instruction using multimedia technology and therefore provide the focus of the remaining discussion of cognitive load theory.

Split-attention effect
Multiple sources of mutually referring information require mental integration that is cognitively taxing. Cognitive load theory suggests that information sources should be integrated when they are unintelligible on their own. When these sources are intelligible on their own, integration is not required. Often, instructional material is presented in both graphic and textual format, with text above, below, or to the side of the diagram.

Cognitive load generated by irrelevant activities (integration of text with diagrams) can impede skill acquisition. Instruction developed by Chandler and Sweller (1991) that integrated disparate sources of information was found to be superior to split-sources of information. Figure 5.5 provides an example of the integrated instruction that provided the basis for these results. Based on this research, Sweller et al. (1990) recommended that, in cases where mental integration is necessary to make sense of two or more sources (elements) of information, integrated instructional formats should replace the conventional approach.

The split-attention effect also occurs when reading conventional experimental reports because the results section occurs before the discussion section, causing the reader to "attend to" two different sections of the paper (Chandler & Sweller, 1992). The nonlinear nature of hypermedia offers a solution to this cross-reference problem because discussion can be hyperlinked directly in the results section. Users then have the option to immediately access the discussion of the results.

Another example of the split-attention effect occurs when learning to use software packages. Learners are instructed to reference multiple sources of information simultaneously: the computer screen, the keyboard, and the computer manual. Cognitively guided manuals that physically integrate the manual, screen, and keyboard were studied without the use of the computer and proved repeatedly to be more effective training tools (Chandler & Sweller, 1996). Split-attention was also reduced in instruction that eliminated the computer and instead used simulation units that placed explanatory text boxes near appropriate keystroke/mouse clicks. Cognitively guided computer instruction was shown to eliminate the need to mentally integrate disparate sources of information, reduce mental load, and enhance learning.

Internal Wiring for Intermediate Switching

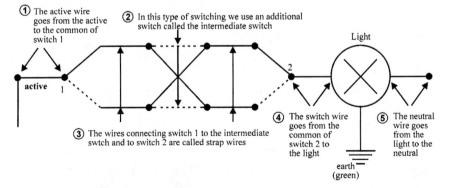

Figure 5.5. Example of integrated instructional format.   Source: Chandler and Sweller, 1991, p. 305.

Instructional materials that split the attention of the learner impose a heavy cognitive load. Schema acquisition cannot commence until the disparate sources of information have been mentally integrated. Material with reduced or unitary sources of information will decrease or eliminate the need for learners to use cognitive resources to structure the material into a form that is suitable for schema acquisition. Learning is enhanced when learners are allowed to attend to schema acquisition rather than to information reformulation (Sweller et al., 1990).

Redundancy effect

Another principle used to reduce cognitive load in instruction is the redundancy effect. Cognitive load theory describes the benefit of providing text only or graphics only displays of information in cases where one medium provides full intelligibility of the concept (Cooper, 1990). Conventional instructional design theory views use of redundant sources as at least neutral and perhaps beneficial in its effect on learning. Conversely, cognitive load theory cautions against the use of simultaneous representations of redundant content in instruction (Chandler & Sweller, 1991). Redundant sources of information place increased demand on cognition that can be freed for processing the intrinsic load. Maps, for example, provide a paper-based graphic instruction that may be completely self-contained. Text used with these maps is redundant and unnecessary. Similarly, many types of text-based instruction (thesaurus, history annals) have no need for graphic support.

In assessing redundant sources of information, it is necessary to accurately evaluate the expertise level of the users and the level of knowledge that they have already mastered. Learning of essential information has been shown to be enhanced by eradicating all nonessential information.

Modality effect

Reduction of extraneous cognitive load caused by split-attention and redundancy effects is one method utilized in developing cognitively guided instruction. Another method that has broad ramifications in multimedia instruction is the modality effect. The modality effect attempts to increase the capacity of working memory by presenting information (instruction) using both aural and visual sensory modes. Research has shown that some portion of working memory is dedicated to attending only to aural information (especially verbal information), and some portion to visual information (especially diagrammatic information). However, as shown previously in figure 5.2, the majority of working memory appears to be in the form of a central resource that serves as the attention-controlling mechanism (Baddeley, 1992).

Cognitive load theory suggests that instruction communicated in dual information modes will expand working memory and thereby enhance performance. An extensive amount of research has been conducted in support of this theory (Penney, 1989). It was found that people were better able to carry out two tasks if the two tasks involved different modalities rather than the same modality. Other research indicated that more items were recalled in a memory test if some of the items were presented in a visual modality and some in an auditory modality, rather than all in a single modality. This research suggested that more memory capacity was available when dual modalities were used. Further, the contiguity principle developed by Mayer and Anderson (1991) states that animation and associated narration are most effective when presented simultaneously rather than sequentially.

Geometry instruction was developed to assess the cognitive impact of the modality effect (Mousavi, Low, & Sweller, 1995). The research used a geometric theory to demonstrate simultaneous processing of narrative descriptions of angles with the figure. This research elected to enhance learning by expanding working memory capacity using dual modality instruction delivery rather than by reducing the split-attention effect by incorporating the text with the graphic. The research found that the dual-mode instructional technique was superior to the visual only format. Although dual-modality involves a split-attention effect, it was shown to supply an effectively larger working memory for learners to assimilate instructional material (Chandler, 1995).

Other studies indicated that simulations that combined audio input with animated flashing, highlighting, or simple movement reduced the search process for learners and enhanced the learning process. However, further research determined that, in instruction where no or little screen search is involved, basic animation may distract or misdirect attention and interfere with learning (Jeung, Chandler, & Sweller, 1997). It was concluded that the additional memory capacity provided in mixed mode presentations was useful only if extensive cognitive resources were not required to relate the audio and visual components of the instruction.

Further studies by Sweller et al. (1990) showed that audio-visual presentations are unlikely to be beneficial if the auditory component is structured in such a way that it overloads working memory (Tindall-Ford, Chandler, & Sweller, 1997). If, for instance, the audio is too long, it will exceed working memory limitations and the instruction will be ineffective. In instances where the audio is highly complex ("high in element interactivity"), it will create excessive demands on working memory. In such cases, Sweller et al. recommend that a written presentation, acting as external memory, will allow repeated scans of related elements and can ameliorate working memory limitations.

A later study revealed that audio-visual instruction may not be ideal when the audio and visual components are redundant or unnecessary for understanding (Kalyuga, Chandler, & Sweller, 1999). Cognitive load theory suggests that if a visual form of instruction is intelligible on its own, supplemental audio streaming is not necessary and can result in cognitive overload. However, for complex instruction that requires an audio component, Sweller et al. suggest that redundant presentation of text should be avoided. Nielsen (1995) cautioned that unconstrained use of multimedia can result in user interfaces that confuse users and make it more difficult for them to understand the information.

Increasingly powerful computer technology has spawned a new methodology—information visualization. The objective of this field is to amplify cognition using computer-supported, interactive, and visual representations of abstract data. These perceptual representations of data have broad implications for reducing cognitive demands of instruction. Research by Mirel (1998) demonstrated the need to develop effective interfaces and cueing for manipulation of graphic data.

These findings have direct implications for the wide range of multi-media technology that is currently available for generating computer-based and web-based instruction.

Applying cognitive load learning theory to the design of web-based instruction

An interdisciplinary team at IIT applied the principles of cognitive load theory to the design of a web-based instructional module on "The Basics of Internet Protocol (IP) over Synchronous Optical Network (Sonet)." The project work plan, described in detail in this section, includes the following: identification of the key elements of the instructional module, design of the graphical user interface (GUI) for the module, and design of the template for delivering the actual instruction.

Identification of key elements of the instructional module

The instruction developed by the IIT team would expand on a text-based module available as part of a web-based tutorial series offered by the International Engineering Consortium (IEC) for the telecommunications industry. The goal for the new instructional module was to deliver the content in a highly interactive and effective format that employed audio-visual streaming of a lecture on the topic.

The IIT team was provided with a copy of the audio-video of the lecturer and the text-based tutorial. Early efforts were devoted to storyboarding the various formats that the web-based instruction would take. This was done in an attempt to clearly understand the "types" of information that the site would need to house, not the actual content of the site. Discussion of the development of the actual instruction follow the discussion of the development of the graphical user interface (GUI).

Design of the graphical user interface

For the purposes of this project, the graphical user interface (GUI) was considered "extraneous cognitive load." That is, it is information that is external to the actual learning content. For these reasons, cognitive load theory would suggest that the GUI should not impose a high cognitive load that will interfere with the learning process. The GUI was developed according to current practice in user interface–web design that stresses functionality and ease of use. Following these design principles should produce an effective GUI that facilitates the learning process without imposing extraneous cognitive load. Figure 5.6 shows the final design of the web-based instruction homepage.

Navigation system

According to Don Norman (1986), one of the original proponents of user-centered design theory, good design involves the development of conceptual models or mental models of a system that allow users to understand it, predict the effects of their action, and interpret their results. Design and development of clear navigational systems improve the usability of a web site by providing a clear conceptual model for the users. Before you can develop effective navigation tools, you must first understand the organization system of your information. Williams and Tollett (1998) reiterated the need for sound conceptual models and stressed that "the focus of good navigation design is organization, not graphics" (p. 132). The design of a good conceptual framework should accommodate all information types and be flexible enough for continual extension and updating (Sano, 1996, p. 89). A study for the National Center for Atmospheric Research reported that one of the biggest problems with hypertext information is that its organization is obscure and hinders users from finding the information they need (Bevirt, 1996).

According to Williams and Tollett (1998), "the primary navigation system should be kept together in a compact package, either at the top of the page, the bottom or off to the side" (p. 133). Our instruction provides the navigation scheme on the left side in conformance with the recommendation of Marcus (1990) to group icons conceptually so users understand how several of them are related. The three icons at the bottom were given a different look to give the appearance of being "clickable." According to Mountford (1990), interface designers should "ensure that the form of objects should follow their functional requirements" (p. 128). These three buttons represent action items: *Download PDF, Ask the Instructor, Acronym Guide*. The icons are separate from the rest of the palette, which contains links to various topics of the instruction. The links light up and buttons compress with mouse-overs. The IEC logo at the top is clickable and returns the user to the IEC homepage.

The initial visual impression is contained on one screen that is entirely accessible without vertical or horizontal scrolling. The title of the instruction is located horizontally at the top of the page. The navigation bar and title of the unit remain stationary, while the instructional content changes. This grid was designed to focus the user's attention on the changing instructional content. The other frames remained fixed. According to Sano (1996), the organization framework "defines a precedence and hierarchy of information that communicates recognizable levels of information" (p. 88) to the user.

The color scheme that is used is uniform with the IEC home page and will provide some consistency as the user migrates to the web-based instruction section. As reported by Sano (1996), "providing location cues and identifiers for the users helps them from becoming lost in hyperspace" (p. 33). The constant presence of the navigation bar provides a shortcut for users and helps them avoid backtracking to the homepage.

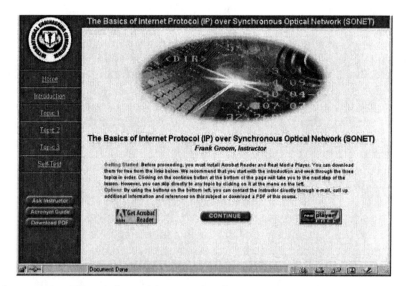

Figure 5.6. IIT Web-based instruction homepage.

Instruction grid
The homepage houses the instruction grid in the center of the page. This is the only segment of the interface that changes; the left portion of the page (menu options) remains fixed. The image is designed to attract the user's attention. It is a layered image of the information industry and creates an aesthetically pleasing and professional design. The course title and instructor name are posted. Users are prompted with *Getting Started* instructions. An introduction to the plug-in requirements and free plug-in links are provided. Users are instructed on the organization of the site, and told how to move through menu selections and to use the "continue" button at the middle-bottom of the instruction grid. Also, users are instructed to use buttons at the lower left of the menu to *Download PDF, Ask the Instructor,* or use the *Acronym Guide.*

With these principles in mind, the GUI was designed with the goal of being extremely intuitive for the user and therefore low in extraneous cognitive load. The actual cognitive load imposed by the GUI was tested in the usability test.

Design of the instruction

As cognitive load theory suggests, the focus of the user's mental resources must be allocated to acquiring the intended knowledge or skill—the intrinsic cognitive load. The purpose of this instruction was to provide a working knowledge of the basics of Internet Protocol (IP) over Synchronous Optical Network (SONET). The course is designed to assist professionals to learn asynchronously about IP.

Storyboarding content of instruction
The IIT team devoted its initial efforts to storyboarding the content and format of the instructional module. The GUI was designed to make the organization of the module visible to the user. It was also designed to accommodate any expansion to the interface.

With the GUI designed, the IPRO focused on the development of the instructional content. The instruction takes place in its entirety in the center of the page. Figure 5.7 shows the templates for the design of the various forms of instruction.

Developing the instruction

Before the instructional content could be designed, it was necessary for the team to decide on the format of the instruction. We viewed the tape and determined that the audio did not match the visuals, leading us to decide between two format options: (1) retape the audio to match the existing visuals, or (2) use the audio and develop new companion visuals. We chose Option 2 because we did not have access to a content expert who could retape the audio portion.

The instructional content would take one of the following three forms:

Streaming Video/Audio Only (figure 5.8). The video/audio streaming was used only in the introduction of the unit. The users were allowed to see the instructor as he introduced himself and the unit. A bulleted list of topics was synchronized with the instructor's introduction.

Audio with Text and/or Diagram (figure 5.9). This format was used to deliver the majority of the instruction. In conformance with cognitive load theory, audio and visuals processed in different streams are the most effective means to deliver highly interactive (complex) material. The nature of this instruction is high in intrinsic load. The delivery of the instruction therefore has to maximize the working memory capacity by presenting the information in dual mode channels (audio-visual).

Removal of the talking head from the screen minimized the split-attention effect and focused the user's attention on the content rather than the instructor. Also, split-attention was minimized in most instances by redesigning the diagrams for clarity, adding bar charts that created visual representations of the information. The follow-on effort included animated diagrams that highlighted areas as they were being discussed to reduce cognitive demand created by screen search. However, as Nielsen (1995) recommended, animations cannot move permanently because anything moving in the user's peripheral vision demands attention.

According to Nielsen (1997), presentation of information on the web must be "scannable." Users will not read dense text; it must be designed to accommodate scanning and quick reading. Text must be visually structured and concise (half the word count of conventional writing). The PowerPoint text slides that accompanied the unit provided the visual medium to ameliorate the demands placed on working memory by long and complex audio streams.

Text only (figure 5.10). The text only function was reserved for the self-test. Users were presented with a series of seven questions. They were instructed to type in the number of the correct answer and press "enter." The computer would then indicate whether the answer was correct or incorrect.

Interactive features

Cognitive load theory recommends expanding the limited capacity of working memory by incorporating dual (audio and visual) modality in the instruction. In an effort to further enhance learning, we have

Streaming Audio/Video	Streaming Audio Only	Text/Graphics Only
Text/Graphics	Text/Graphics	
Design I	**Design II**	**Design III**

Figure 5.7. Templates for the design of the various forms of instruction.

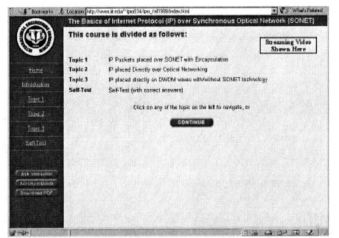

Figure 5.8. Streaming audio/video template for instruction.

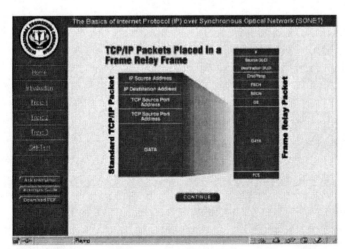

Figure 5.9. Streaming audio with text and/or diagram template for instruction.

introduced multiple (audio, visual, and tactile) modality in our instruction by including the following interactive components:

Ask the Instructor	Clicking on icon brings up e-mail screen to e-mail the instructor directly with questions.
Acronym Guide	Contains supplemental material for instruction including acronym guide and reference links.
Download PDF	Provides hardcopy of instruction and any other material that the user needs for effective instruction.
Self-Test	Offers opportunity for users to test their mastery of the instruction. Students can then revisit sections of the module as necessary. Correct answers are shown after each question to reduce the split-attention effect caused by delayed answers to questions.
Online Evaluation	Provides a tool at the end of the instruction to gather user feedback (on the instruction and the instructor) so that the site can continuously be revised.

The learning effectiveness of the instruction was tested in a usability test described in the next section.

Designing and conducting a usability test of the web-based instruction

John Gould (1988) provided four guidelines in designing usable products: early focus on users, empirical measurements, iterative design, and integrated design. Conducting usability tests is a critical method in accomplishing these objectives.

Usability testing laboratories are designed to conduct tests that assess the human interaction with training–instructional materials and the effectiveness of the human interface embedded in various products and software. Such testing is designed to reveal the inadequacies of the product interface and to provide immediate feedback for iterative product design. Usability testing laboratories provide a user-centered focus that is critical to the design of successful products.

Usability testing laboratory and evaluation center

The usability test was conducted in the Usability Testing and Evaluation Center (UTEC) designed by Dr. Susan Feinberg, UTEC Director and Professor of English at IIT. The lab consists of a user observation room and a testing room connected with a one-way mirror. State-of-the-art

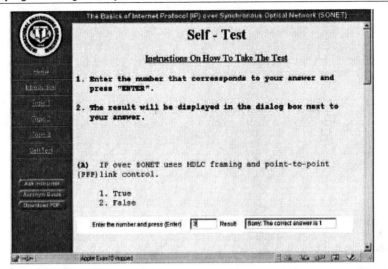

Figure 5.10. Text only template for instruction.

equipment includes a PC/LCD monitor, 2 video cameras with tripods, 2 monitors, a mixer, and a VHS recorder/player. Users sit in front of the PC to test software or a web site. The results of the user's interaction with the computer interface (cursor movement, keystrokes, screen shots) are fed directly to the monitor in the observation room (figure 5.11).

The cameras can be focused on specific components of the interface that evaluators are interested in observing. For example, the cameras can display the user's physical reaction to the interface (facial expression, body position) or the keyboard/mouse to observe keystrokes, etc. Monitors allow simultaneous split-screen viewing of all images with selected placement of images on the screen.

Developing the usability test

According to the guidelines for usability testing outlined by Dumas and Redish (1994) and Rubin (1994), usability testers must design a test that probes areas of potential usability problems. In general, tasks should be performed in the natural order that the user will encounter them. Evaluators must also design parameters for measuring the success of the interaction. These parameters can be performance measures that include counts of actions and behaviors that you can see (time to finish a task, number of wrong menu choices, number of errors, number of attempts at a task). Parameters can also be subjective measures including the user's perceptions, options, and judgments. Subjective measures can include writing the comments of the user or asking users to complete a questionnaire rating the effectiveness of various features of the software/ product. Evaluators create a scenario that sets the stage for the user and makes the test situation less artificial. Users must be given a list of required tasks that are clearly delineated at the start of the test.

Our usability test consisted of the following components:

Participant Profile   A participant profile was developed and completed by each participant to document the background of the user. Participants must represent the group that will actually use the product. In this test, users were selected from an electrical engineering or telecommunications background.

Participant Recruitment   Virzi (1992) found that 80% of the usability problems in a product were detected with between 4 and 5 participants and 90% were detected with 10 participants. All of the global problems were detected with 10 participants. A typical test includes 6 to12 participants. Our usability test used 6 participants.

Consent Form   A consent form was prepared as a legal document informing the user of the procedures involved in the test, explaining risk factors, and providing information on the data retention and anonymity of the users. The signatures of the principal investigator and the user were obtained and dated.

Figure 5.11. Usability testing laboratory set-up.

Script	A script was prepared that described the instruction, gave directions to the user, and described the test procedure. This information was read from the script to ensure that all users received the same introduction.
Task List	A task list was prepared to test the usability of the web interface and the content of the instruction. The self-test in the instruction provided an objective assessment of the effectiveness of the instruction.
Debriefing Questionnaire	A debriefing questionnaire provided a numerical ranking of the user's perception of effectiveness of the instruction and the user's attitude toward the web site and learning via this medium.

Recommendations for redesign of the web-based instruction

The following are the recommendations for the redesign of the web-based instruction based on cognitive load theory, results of the usability test, and user responses to the posttest questionnaire. All six users completed the usability test.

Graphical user interface

- *Include Topic Titles.* Titles of topics should be included with their menu entries (either as mouseovers, pop-ups, or link names) so that users know the subject matter that is contained in each topic.
- *Improve Acronym Guide.* Make it clear to the user that the *Acronym Guide* exists. Acronyms in the instruction should also be linked directly to their entry in the guide so users can click for immediate definition.
- *Download PDF.* Users should be instructed on the home page that they might want to download the PDF file before they begin the instruction. This could also be stated in an introductory demo of the unit.

Overall, the users interacted effectively with the graphical user interface and, on average, ranked its design as "above average."

Self-test

- *Include radio button feature.* The self-test should be redesigned to eliminate the need for the user to type a number and enter it for each answer. Rather, the user should click on a radio button located next to the selected answer. Such a design will eliminate the cognitive load caused by unnecessary screen search and the split-attention effect.

- *Check test validity.* Four of six users missed the same three test questions. Instructional designers must ensure that the concept that is being tested is covered adequately in the instruction. Word questions clearly; eliminate double negatives in test questions. Such revisions allow the user to focus on the content that is being tested.
- *Place questions near content.* When material is difficult (high in intrinsic cognitive load), it may be better to place the test questions immediately after the learned content instead of at the end of the instruction. Such placement eliminates the cognitive load imposed by all material presented after the tested content has been presented. A cumulative test of all concepts may be provided at the end of the module.
- *Provide answer feedback.* User should receive an explanation after each test response indicating why the answer is right or wrong. Response could also refer the user to a certain section or provide a link to additional reference or resource material. Such layering of information provides the user the opportunity to explore these options if necessary, or to continue the test.

Instructional design

- *Upgrade audio quality.* The quality of the audio was choppy and garbled in some places. The instructor was reported by some users to be too slow, by some to be too fast when difficult material was being discussed. The audio must be retaped at a speed that coincides with the presentation of the material. Because this is the primary delivery mode for the content, much consideration must be given to its quality and speed of presentation. If not, audio presentation of the content is no better than straight visual presentation, and may even impose greater cognitive demands than straight visual presentation.
- *Include audio control.* Users commented on the need for an audio control panel so they could pause the audio to review material, take notes, etc. Others suggested the need to turn off the sound entirely when they are revisiting portions of the instruction. Some users also wanted a gauge of how much audio remained on the page. All of these options provide user control, a critical principle of web-based design, and reduce extraneous cognitive load.
- *Animate graphics.* The instruction content is high in intrinsic load; the companion graphics are also difficult to understand.

Cognitive load theory suggests that relevant areas of a complex figure should be highlighted as they are being discussed in the audio. This eliminates cognitive load induced by the split-attention effect introduced by audio playing over complex graphics that require a high search component. Moreover, additional information can be provided through closer view of certain areas of the graphic or links to additional information in the reference guide.

Redesign of instruction based on recommendations from usability test

The first usability test led to recommendations that we employed in our second design of web-based instruction. These recommendations incorporate the principles of cognitive load as we applied learning theory techniques to the redesign of content and the interface.

Instructional design

Our users for the first test were critical of the audio presentation of the instruction. Their problems and recommendations were supported by the principles of learning theory. Users complained that the audio was garbled, too fast, too slow, and everything in-between. After we improved the quality and speed of our audio delivery, we gave the users audio control over the instruction in the form of a slide that allowed them to pause, repeat, or continue the audio portion of the instruction. This control reduced split-attention that interfered with the intrinsic content of the material and increased modality by allowing the users to have an interactivity with the screen (touch mode), thus enhancing working memory. The main lesson here is that users wanted more control over content and interactivity with the mode of delivery (audio).

Finally, as a result of the first user test, we redesigned the graphics, again incorporating the techniques to improve instruction based on cognitive load theory. In our complex graphics, we eliminated screen search, split-attention, and redundancy by highlighting and animating graphics. With audio instruction to accompany visual instruction (so that multimodality would increase working memory), we used a single graphic that visually built content and highlighted the information as the audio information (another modality) provided additional context. Figure 5.12 shows this instruction at three distinct points in the sequence of instruction.

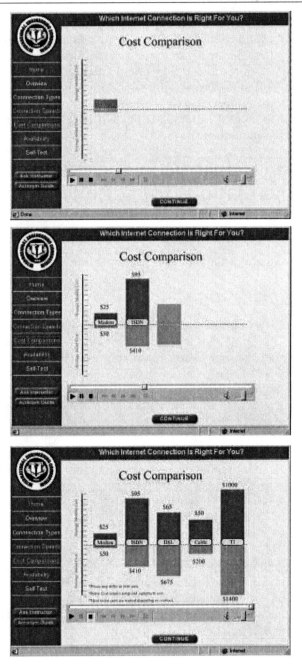

Figure 5.12. Redesigned graphics use multimedia delivery to reduce cognitive load; audio supplements visual content.

Conclusion

The objective of this research was to examine the basis of cognitive load theory and determine its applicability to the design of web-based instruction. The basic premise of cognitive load theory is that the focus of an instructional module must be the instruction itself (intrinsic cognitive load). Information that is adjunct to the instruction (extraneous cognitive load) must be designed to reduce cognitive demand. Cognitive load theory provides general design principles to reduce extraneous cognitive load caused by the format of the instruction and to increase the limited capacity of working memory.

The following are the central conclusions of this research:

1. Cognitive load provides fundamental learning theory for the design of effective web-based instruction. It forces the instructional designer to distinguish between intrinsic cognitive load (actual instruction) and extraneous cognitive load (information that is extraneous to learning). Such distinctions provide a necessary structure for effective design.
2. The principles of split-attention, redundancy, and modality identified in cognitive load theory have clear applications in the design of web-based instruction.
3. Cognitive load theory is consistent with general web design principles and provides a learning-based criterion for effective design of web-based instruction. In considering cognitive demand of web-based instruction, the graphical user interface is considered extraneous information. Therefore, the interface must be designed to give the user clear access to the instruction without imposing unnecessary cognitive demand.
4. Multimedia instructional formats must be developed in consideration of cognitive load principles to effectively increase working memory capacity.

A checklist for applying cognitive load to web-based instruction

The guidelines on this page can be used for web-based instruction that incorporates the principles of cognitive load learning theory.

Checklist

Before You Begin

☐ Be sure to include the content expert on the instructional design team.
☐ Clearly define the target "learners," their level of expertise and the purpose of the instruction.

Developing the GUI

☐ Review content, classify types of instruction and develop storyboards for each type of instruction.
☐ *Reduce extraneous load*: Make sure the web interface that houses the instruction is intuitive, easily navigable and well organized.

Developing the Instructional Content

☐ Distinguish between intrinsic and extraneous cognitive load components.
☐ Eliminate split-attention and redundancy effects from instruction.
☐ Present simple content in simple forms with few interactive elements.
☐ *Increase working memory capacity*: Use multiple modalities to present highly complex, highly interactive material.
☐ Use highlighting to eliminate screen search from complex graphics.
☐ User test instruction to determine the validity of the content and the effectiveness of the interface and the instruction, and to assess the accuracy the learner profile.
☐ Revise instruction to incorporate user feedback.

References

Baddeley, A. (1992). Working memory. *Science, 255*, 556-559.

Baecker, R., Grudin, J., Buxton, W., & Greenberg, S. (1995). *Readings in human-computer interaction: Toward the year 2000.* San Francisco: Morgan Kaufmann.

Bevirt, B. (1996). Designing hypertext navigation tools. Digital Information Group, National Center for Atmospheric Research web site: www.sch/ucar/dig/brainb/design/navtool.p.html.

Chandler, P., (1995). Is conventional computer instruction ineffective for learning? www.penta2.ufrgs.br/edu/telelab/6/chandler.htm.

Chandler, P., & Sweller, J. (1991). Cognitive load theory and the format of instruction. *Cognition and Instruction, 8*(4), 293-332.

Chandler, P., & Sweller, J. (1992). The split attention effect as a factor in the design of instruction. *British Journal of Education Psychology, 62,* 233-246.

Chandler, P., & Sweller, J. (1996). Cognitive load while learning to use a computer program. *Applied Cognitive Psychology, 10,* 151-170.

Cooper, G. (1990). Cognitive load theory as an aid for instructional design. *Australian Journal of Educational Technology, 6*(2), 108-113.

Cooper, G. (1998). Research into cognitive load theory and instructional design at the University of New South Wales. www.arts.unsw.edu.au/education/CLT NeT Aug 97.html.

Dumas, J., & Redish, J. (1994). *A practical guide to usability testing.* Norwood, NJ: Ablex.

Gould, J. (1988). How to design usable systems. In M. Helander, (Ed.), *Handbook of Human-Computer Interaction* (pp. 757-789). North-Holland: Elsevier.

Jeung, H., Chandler, P., & Sweller, J. (1997). The role of visual indicators in dual sensory mode instruction. *Educational Psychology, 17*(3), 329-343.

Kalyuga, S, Chandler, P., & Sweller, J. (1999). Managing split-attention and redundancy inmultimedia instruction. *Applied Cognitive Psychology, 13,* 351-371.

Marcus, A. (1990). Principles of effective visual communication for graphical user interfacedesign. Excerpt from Baecker et al., *HCI: 2000,* 425-441.

Mayer, R. & Anderson, R. (1991). Animations need narrations: An experimental test of a dual-coding hypothesis. *Journal of Educational Psychology, 83,* 484-490.

Miller, G. A. (1956). The magical number seven plus or minus two: Some limits on our capacity for processing information. *Psychological Review, 63,* 81-97.

Mirel, B. (1998). Visualizations for data exploration and analysis: A critical review of usability research. *Technical Communication, 45*(4), 491-509.

Mountford, S. J. (1990). Tools and techniques for creative design, Excerpt from Baecker et al., *HCI: 2000*, 128-141.

Mousavi, S., Low, R., & Sweller, J. (1995). Reducing cognitive load by mixing auditory and visual presentation modes. *Journal of Educational Psychology, 87*, 319-334.

Nielsen, J. (1995). Guidelines for multimedia on the web. Jakob Nielsen's *Alertbox, www.useit.com/alertbox/9512.html.*

Nielsen, J. (1997). How users read on the web. Jakob Nielsen's Alertbox, *www.useit.com/alertbox/9710.html.*

Norman, D. (1986). Cognitive engineering. In D. A. Norman and S. W. Draper, *User-centered system design: New perspectives in human-computer interaction.* Hillsdale, NJ: Lawrence Erlbaum Associates.

Penney, C. (1989). Modality effects and the structure of short-term verbal memory. *Memory and Cognition, 17*, 398-422.

Rubin, J. (1994) *Handbook of usability testing. How to plan, design, and conduct effective tests.* New York: Wiley.

Sano, D. (1996). *Designing large-scale web sites: A visual methodology,* New York: Wiley.

Simon, H. (1974). How big is a chunk? *Science, 183*, 482-488.

Sweller, J., Chandler, P., Tierney, P., & Cooper, G. (1990). Cognitive load as a factor in the structuring of technical material. *Journal of Experimental Psychology: General, 119*, 176-192.

Sweller, J. (1994). Cognitive load theory, learning difficulty and instructional design. *Learning and Instruction, 4*, 295-312.

Tindall-Ford, S., Chandler, P., & Sweller, J. (1997). When two sensory modes are better than one. *Journal of Experimental Psychology: Applied, 3*(4), 257-287.

Virzi, R. (1992). Refining the test phase of usability evaluation: How many subjects is enough? *Human Factors, 34*(4), 457-468.

Williams, R., & Tollett, J. (1998). *The non-designer's design book.* Berkeley, CA: Peachpit Press.

6

What Makes Up
A Procedure?

Hans van der Meij
Peter Blijleven
Leanne Jansen
Twente University

Procedures play a vital role in technical documentation as they guide people in performing a task. They are the heart of most manuals. It is therefore somewhat surprising that the theoretical and empirical knowledge on their nature has remained somewhat elusive. This chapter advances a theoretical framework for creating and describing procedures. In addition, it illustrates and tests this framework with data from 52 software manuals and 52 hardware manuals. Our goals for the study were similar to the ones identified in Brockmann's (1998) exploration in the history of technical communication in the United States. That is, we set out to create a better sense of self-identity and tradition and in doing so better position ourselves for perceiving the strengths and weaknesses of existing practices.

The task proved to be difficult. We struggled to generate a state-of-the-art description that does justice to theory as well as practice. For example, early in the process we had to give up on the idea that we might discover a limited number of clearly identifiable genres because we found too much textual and visual variation. Insights came in bit by bit. One of the main problems, as we discovered, was striking a good balance between the design or plot of a procedure and its details.

The publication of a taxonomy of procedural discourse by David Farkas (1999) helped us create an enlightening domain-related organization. The Four Components model of a procedure that we advance here suggests that procedures are basically made up of the following components: goals, prerequisite states, unwanted states (warnings and problem-solving information), and actions & reactions.

The chapter is structured as follows. First, we discuss the states and actions model that underlies our Four Components model. Thereafter, we present the method detailing how the procedures were sampled and analyzed. The following sections then describe each of the four components. All these sections contain a theoretical description, a set of design guidelines, and a discussion of the findings results from the inventory. We conclude the chapter with a brief statement on how these insights can further be used in the design of technical communication. [Note: For unwanted states, we discuss only warnings. A discussion on problem-solving information can be found in Van der Meij (1999b).]

The logical and rhetorical construction of procedural discourse

Farkas (1999) presented a framework that focuses on the analysis of procedures. The framework is grounded on two theoretical perspectives, namely systems theory and rhetoric. The basic tenet of systems theory is that a procedure consists of a mixture of states and actions. Systems theory offers a logical point of view. Rhetoric places this logical view within a social context. Rhetoric stresses that variations are the rule rather than the exception as designs are adapted to audience and context. Rhetoric also draws attention to the need to establish source credibility ("selling oneself") and product credibility ("selling the domain") and to engaging and persuading the user. We return to rhetoric later; here we briefly sketch systems theory.

According to systems theory, a procedure is a complex mixture of states and actions. A procedure must inform a user about system states and actions that change these states. Because all procedures are assumed to derive from purposeful human behavior they are expected to have an identical underlying logical structure. Farkas (1999) distinguished four system states and three action types (figure 6.1). In various combinations these elements form procedures. For example, a complete procedure informs users about a desired state or goal, outlines the conditions for action, presents intermediate states, and helps the user prevent and overcome problems. In addition, the user is told which actions to take to reach the goal, how the system is likely to respond, and what else may happen.

Systems theory led Farkas (1999) to the construction of a prototypical or "streamlined-step" procedure. The streamlined-step procedure consists of five components: a title that introduces and briefly explains

Desired state:	The goal presented to the user.
Prerequisite state:	The condition for moving toward the desired state.
Interim state:	The intermediate state or subgoal.
Unwanted state:	The to-be-avoided states (e.g., errors, malfunctions).
Human actions:	The actions taken by the user.
System actions:	The responses of the system.
External actions:	The events or actions from outside (e.g. power shortage) that may effect the system.

Figure 6.1. The system states and actions that constitute a procedure.

the goal of a procedure; a conceptual element that elaborates the goal; an infinitive subheading that clarifies the purpose of subgoals; steps with action statements and system responses; and notes that present information that lies outside the main flow of the procedure (e.g., warnings). The streamlined-step procedure is prototypical because it contains clearly identifiable and more or less chronological components. In addition, its brevity and simplicity make such a procedure easy for the user to process.

Following the description of method, we present our own streamlined version of procedures in the remainder of this chapter. More specifically, we discuss our Four Components model. The components in this model are goals, prerequisites, actions & reactions, and unwanted states. In the latter component we make a distinction between warnings and problem-solving information. Befitting their general presentation order, warnings are discussed before the actions & reaction component and problem-solving information after that component. In discussing each component we advance pertinent theoretical notions and offer recommendations for practice in the form of design guidelines. In addition, an inventory section describes the findings for many of these guidelines and illustrates some of the important design variations that we found for each component.

Method

The sample contained 52 software manuals and 52 hardware manuals. All were published between 1991–1998. The mean publication date is 1995. Only publicly available or commercial documents were sampled. The manuals came from the maker of the product, or they came from third parties such as publishers. For the study a codebook was created, tested, and revised until there was satisfactory agreement between observers in coding the data. Sampling and scoring of the data was also based on this codebook (Van der Meij, Blijleven, & Jansen, 2001).

Three main types of manuals were distinguished: tutorials, reference guides, user guides. Between user guides a distinction was made between the tutorial part of the guide and the reference part. Tutorials are defined as manuals for novices and beginners. They teach users how to perform a handful of basic tasks using the product. Reference guides are manuals for intermediates and expert users. They give more conceptual inforamtion than tutorials and generally describe all the possibilities of a product. User guides are often hybrids, serving both as a tutorial and as a reference guide. In general they often have a clearly delineated section that serves as a tutorial and a section with reference information.

From the manuals we also recorded size, content, accessibility, and reader guidance. Size and content are simply the number of pages of the book and the type of program or apparatus involved. To assess accessibility we looked at the index of the manual. Indexers generally recommend the presence of a sizeable index to improve the accessibility of a document. For technical documents such as a software or hardware manual, the size of an index should be about 5 % of the total number of pages of the book (Brockman, 1990; Lathrop, 2001; University of Chicago Press, 1982). This criterion of 5% typically refers to indexes with a two-column design. For reader guidance we examined whether the manuals used symbols or icons to alert the user to different information types and explained these in the front matter of the book, as designers are advised to do (Price & Korman, 1993). For example, when a manual presents a flashing bulb icon to demarcate the presence of a tip, the reader should be informed about this convention.

The sampled software manuals

The mean size of the 52 software manuals is 411 pages with a standard deviation of 330 pages and a range of 18 to 1,344 pages. The sample consisted of 17 tutorials, 16 reference guides, and 19 user guides (7 tutorial part, 12 reference part). The distribution for type of software was: 12 manuals for operating systems (e.g., MS-DOS 5.0, Windows 3.1, Windows95, Windows98, and Macintosh OS), 9 manuals for general purpose software (sound-edit software, games, scheduling software, presentation software, and Norton commander), 9 manuals for communication software (Eudora, Netscape Navigator, internet provider, fax, html & internet, and network-surfing), 7 manuals for word processing software (e.g., FrameMaker, Word, and WordPerfect), 4 manuals for spreadsheets (Harvard graphics and Lotus 1-2-3), 3 manuals for databases (e.g., Access '95 and '97), 5 manuals for graphical software (e.g., Corel Draw, Print artist, Coral magic, and Iomega MultiMedia) and 3 manuals for programming languages (Visual Basic and HTML).

Thirteen (25%) software manuals were not indexed, ignoring an important means of accessing information. These manuals had a mean size of 125 pages. One of these manuals even consisted of 568 pages. Thirty-nine (75%) software manuals were indexed. Seventy-seven percent of the indexes in these manuals had a two-column design. The mean size of the indexes was 3.1%. Only three indexes scored 5% or higher. These results are comparable to the findings from an earlier study of software manuals (Van der Meij, 1996a)

Thirteen (25%) software manuals gave reader guidance in the form of symbols or icons. Eight of these manuals failed to explain this convention in the front matter of the book.

The sampled hardware manuals

The mean size of the 52 hardware manuals is 40 pages with a standard deviation of 38.4 pages and a range of 5 to 202 pages. The sample consisted of 1 tutorial, 6 reference guides, and 45 user-guides (7 tutorial part, 38 reference part). The distribution for type of hardware was: 9 manuals for video cards, 8 manuals for CD-ROM players, 8 manuals for monitors, 7 manuals for printers, 7 manuals for motherboards, 4 manuals for soundcards, 2 manuals for mice, 3 manuals for modems, and 4 manuals for other hardware (ZipDrive, network card, joystick, scanner).

Forty-five (87%) hardware manuals were not indexed. This is comprehensible considering the size of most of these manuals. Thirty-two non-indexed manuals were thin booklets with an average number of 30 pages (mean=29.5, standard deviation=33.5, maximum=40). For the other non-indexed manuals accessibility is probably difficult, especially for the manual with 202 pages. Seven hardware manuals were indexed. Seventy-one percent of these indexes had a two-column design. The mean size of the indexes was 4.2% (range 2%-7%), which is close to the recommended criterion.

Five hardware manuals (10%) provided reader guidance in the form of symbols or icons. Three did not explain this convention to the reader.

Goals

A goal is a state that the user tries to realize. In terms of system theory this may be either a desired state or an intermediate one. Goal sections may include a description explaining what the goal is and why the user should pursue it ("selling the domain"). When there is more than one method to achieve the same goal the designer must decide whether to present these alternative methods. When two or more alternatives are presented users also need information about the proper conditions for selecting one over the other. We discuss these issues under the rubric of task analysis and GOMS.

Goals should fit the user's skill and information need and should make sense to the user. This is often codified in the (sub) title of a procedure. Titles and subtitles can thus serve an important function in conveying the big picture of the tasks that are involved in the use of a program or apparatus. We discuss various issues involved in creating meaningful goals and titles

Task analysis and GOMS

As procedures guide people in performing a task, an important action of designers is performing a task analysis. An influential model for such analysis in the field of human computer interaction is GOMS (Card, Moran, & Newell, 1983). The foundation for GOMS is the same as for the logical model, namely the (bounded) rationality principle of Newell and Simon (1972). According to this principle people act so as to attain their goal through rational or logical action, but they may fail to achieve their goal due to cognitive limitations. The principle has been summarized in the following formula: "Goals + Task + Operators + Inputs + Knowledge + Processing Limits ≈ Behavior" (Card et al., p. 27). The first three factors lend themselves more readily to an advanced analysis than the latter three which are situation-specific. The GOMS model is therefore summarized as "Goals + Task + Operator ≈ Behavior."

The factors "Goals + Task + Operators" represent the logical situation. They lend themselves very well for an objective analysis of the structure of tasks. *Goals* embody the intention of the user. They may range from high-level ones such as "creating a style guide" to low-level goals such as "typing a character." Goals also include what Farkas (1999) called the desirable state and the intermediate state. The *task* component is split up in two distinct factors, namely method and selection rules. In most tasks the user has to execute a number of consecutive actions. Such a sequence of actions is called a method for achieving a goal. When there is a choice between two or more methods for achieving the same goal, the user needs information for selecting the most appropriate one. *Operators* represent the observable user actions such as touching the screen, pressing a key, or clicking the mouse. Together, the components Goals, Operators, Methods and Selection rules form the GOMS model.

GOMS is a strong model for task analysis in the field of human computer interaction, which is why Duffy, Palmer and Mehlenbacher (1992) use it as an aid to develop content for their design of online help. One advantage of GOMS is that it forces the designer to select that particular level of operators that is most suited for the audience. In other words, a GOMS analysis reveals which assumptions a designer holds of the entry-level skills of the audience for a manual. For example, if a GOMS task description includes the following operator "inserting a picture" it would stop there because operators are the base level of analysis. The assumption is then that the users already knows which picture to insert, how to select a picture, how to position the cursor, and which menu options to choose. Rhetoric plays an important role in making such entry-level choices, more notably the concept of audience adaptations.

GOMS also draws attention to the fact that a task description of a procedure may involve a choice of method and may mean that the user should be informed about selecting the most appropriate one. In many situations there is more than a single method to achieve a goal. The designer must therefore decide whether to present one or more of these methods and, in the latter case, in what order to do so. For example, in software programs tasks can often be achieved by using the mouse in a menu-based method or by using the keyboard in a shortcut version. Because the first is more intuitive, it is the preferred option for first-time users of a program.

For example, in word processing programs the task "text deletion" can be achieved by pressing delete or backspace, by selecting menu options such as delete or cut, by not saving a file, and so on. The efficiency of these methods varies with the context. Minimalism suggests that context should therefore play an important role in instructing the user about these different methods (Van der Meij, 1992; Van der Meij & Carroll, 1998). That is, the value of a particular method of text deletion ought to be self-evident from the type of deletion task that the user should complete in a procedure. Thus, the use of the Delete key is practiced in correcting small typos in a text and the use of a menu option for deletion is linked to deleting one or more paragraphs of text.

GOMS offers an excellent view on rational human information processing. But an expansion of the factors unwanted states and explorations is needed to make it a more comprehensive model for procedures. Preventing or handling *unwanted states* is ignored in GOMS whose original purpose was to predict the execution times of flawless performances. User *explorations* are ignored because these actions tend to serve fuzzier goals, at least initially, than can be fitted within a rational view. Both factors are treated in detail later on in this chapter. Unwanted states are a distinct component in our model of procedures. User explorations are discussed in the component action & reaction.

Creating meaningful goals and titles

We once came across a procedure in which the user had to complete 21 action steps to achieve a task. There was no breakdown into subgoals and no information about the subgoals that the user was evidently completing. Also there was very little feedback. There was no explicit information about task progression and only twice in the whole procedure did a screen capture provide a point of reference. In short, all users frequently saw was "Click OK, Input x, Click OK, Click OK again" etc. Such a procedure clearly needs to be improved.

One of the fixes should be in presenting subgoals, because at some moment during the 21 action steps users are likely to question the goal they are pursuing. To break down a goal into subgoals, the designer could decide to make chunks of about 5 to 7 action steps. This rule of thumb refers to a conservative derivative from the famous rule of seven plus or minus two which stands for the number of items that adults can keep in short-term memory. When the user has to memorize the action steps, the advice of creating (sub)goals with 5 to 7 action steps can be useful.

The designer can also opt for a breakdown that is based on the domain at hand. One of the choices may stem from the task analysis. Invariably such an analysis shows that users frequently reuse certain methods. These repetitive methods present logical subgoals. For example, in word processing one recurring method involves manipulating text blocks. Another option is to invent meaningful (sub)goals, to give tasks a meaning that they do not have by themselves. For example, users who must learn how to use a mouse are not interested in learning these movements themselves. They are a means to an end. Therefore, it is useful to couch the practice in tasks that are (more) meaningful, such as having users engage in tasks involving the use of an electronic calculator. In the programming domain it is likewise not obvious what the core tasks are. For this reason, Carroll and his colleagues created applications and asked the users to debug these. In one of these games they presented a special "stealMove" that permitted the game to take two moves at once occasionally. In addition to being a manageable task it was also a representative programming task.

Titles can help the user keep in view the big picture of the skills being learned. In addition to being task oriented, titles should therefore convey the major tasks of the user. For example, titles such as " Editing and entering text," "Formatting text," "Spell-checking text," and "Finding and changing items in a document" in a word processing manual clearly reflect some of the major structural components for this type of software. Titles that reflect the task structure may support users by offering scenarios and by supporting the different points of view users should take in task execution. In addition, they can help users locate information easily when the manual is consulted for reference.

Farkas (1999) indicated there are four main ways to phrase titles, namely as: noun phrase (e.g., button duplication), gerund (e.g., duplicating buttons), root (e.g., duplicate buttons), and infinitive (e.g., to duplicate buttons). According to Farkas noun phrases are the least informative and least often used. Gerunds are the "classic" choice because they convey a sense of process and work well over a broad range of designs. Roots have the drawback of sounding like directives. The preposition "to" in the infinitives prompts a "title-to-step" design. Infinitives can be effective as subtitles for closely related subgoals that should be clustered.

Recommendations for practice

Design guidelines for the Goals component are:

- The title of a procedure should be task-oriented.
- The title of a procedure should reflect the task structure.
- The title of a procedure should be in gerund form.
- The title of a procedure should present the most general action leading to the goal state.
- A goal description should sell the domain.
- Inform users of the selection rules when there are different methods to achieve the same goal.
- For introductory usage, present only the most intuitive (e.g., menu-based) method early in the documentation.
- When it is important that users memorize a procedure it is preferable to cluster the information so that each (sub)goal asks the user to perform maximally about 5 to 7 actions.

Results from the inventory

On task analysis and GOMS

Tables 6.1a and 6.1b show how often users receive information about alternative methods, different ways to achieve the same goal. As the tables indicate, alternative methods do not abound. About 30% of the software manuals and 10% of the hardware manuals present an alternative solution in their procedure. When alternatives are presented, they are typically given within an action step. Almost 60% (57%) of these invite the user to choose between a menu-based approach or a function key based approach (figures 6.2 and 6.3). Alternative methods can also

Table 6.1a. Taskanalysis and GOMS in software manuals (n=52). Do manuals present alternative methods and selection rules?

	N	%
Alternative action steps	11	21.2
Alternative approaches	5	9.6
Presence of selection rule	1	1.9

Table 6.1b. Taskanalysis and GOMS in hardware manuals (n=52).

	N	%
Alternative action steps	3	5.8
Alternative approaches	1	1.9
Presence of selection rule	0	0

include alternative approaches in which the user actions extend beyond a single action step. Two illustrative examples of such alternative approaches are given in the figures 6.4 and 6.5. Figure 6.5 is from the oldest manual in the sample. The figure nicely illustrates that the presentation of alternative methods may hinge on the choice of input device.

Selection rules inform users about the advantage(s) of a particular method. There is hardly a need for their presence in the sample because, as indicated earlier, most of the alternatives involve simple shortcuts that are more or less self-evident. Only once were users given selection rule information. Figure 6.6 shows this rare exception. The information is presented at the end of the procedure, presumably to serve as a form of afterthought or tip. It stands out on the page thanks to the little figure and the italicized text.

On creating meaningful goals & titles

Tables 6.2a and 6.2b indicate that the average procedure consists of four action steps which tend not to be split further. Subgoals were distinguished only occasionally. This suggests that the designs are just about right, because the mean scores are well below a criterion of 5 to 7 action steps. But unfortunately one cannot draw such a conclusion from just these data. More detailed analyses are needed to find out whether designers adhere to the guideline of minimizing the number of user actions in procedures.

1. From the main Harvard Graphics window, select **File I New presentation** or press **Ctrl-N**. The Add Slide dialog box appears.

Figure 6.2. An alternative method in an action step (source: Harrison & Yu, 1992).

3. **Press Enter** or click Add.

Figure 6.3. An alternative method in an action step (source: Gardner & Beatty, 1994).

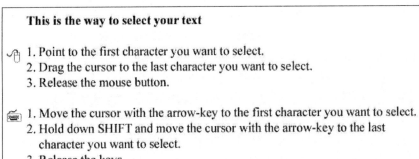

3. Click OK at the bottom of the Data Form to display the Slide Editor view.
4. From the menu bar, select Chart | Chart Options to display the Chart Options dialog box.

TIP

You can access this dialog box directly by pressing F8 from the Slide Editor, then move to the other option menus by clicking the buttons.

Figure 6.4. An alternative method involving more than a single action step (source: Harrison & Yu, 1992).

This is the way to select your text

1. Point to the first character you want to select.
2. Drag the cursor to the last character you want to select.
3. Release the mouse button.

1. Move the cursor with the arrow-key to the first character you want to select.
2. Hold down SHIFT and move the cursor with the arrow-key to the last character you want to select.
3. Release the keys.

Figure 6.5. An alternative method involving more than a single action step (source: Microsoft Corporation, 1991).

Table 6.2a. Meaningful goals and titles in software manuals (n=52).
What are the number of action steps and subgoals in software
manuals and how often are the actions supported by a visual (i.e., a
figure) and by additional textual information?

	Mean	S.D.	Range
Action steps	4.47	2.04	2-12
Subgoals	0.27	0.45	0-1
Separated visual feedback	0.85	1.16	0-6
Separated, additional textual information	0.82	1.36	0-6
Integrated, additional textual information	1.29	1.40	0-7

Table 6.2b. Meaningful goals and titles in hardware manuals (n=52).

	Mean	S.D.	Range
Action steps	4.33	2.64	1-15
Subgoals	0.15	0.36	0-1
Separated visual	0.83	1.15	0-5
Separated, additional textual information	0.58	0.89	0-3
Integrated, additional textual information	0.93	0.99	0-3

If you want to fixate the number notation of the whole worksheet, you should choose Style Worksheet-standars and select the notation in the listbox Numberstyle.

Figure 6.6. An alternative method along with rule information. In
this case the user is attended to the possibility of fixating the data-
input via a single window rather than through several dialog boxes
(source: Evers, 1994).

The difficulty has to do with the precise content of an action step. Some action steps ask the user to perform just one action, but there are also action steps that involve more actions.

Indeed, some action steps even asked the user to perform at least five different actions. In such a case, users may find it hard to track the goal they are pursuing.

Tables 6.2a and 6.2b indicate that bare bone instructions are rare. The action steps often include other information besides an instruction. Users also frequently receive additional textual information such as an explanation, suggestion, and feedback (e.g., "the Dialog Box appears"). Users receive this information at a rate of about once for every two action steps in software manuals and once for every three action steps in hardware manuals. Tables 6.2a and 6.2b indicate that such information is presented more often in integrated form (e.g., as part of the action step) rather than as a separate unit. On average, 66% of all action steps in a software manual is accompanied by additional information. For hardware manuals this lies at 54%. The average manual also presents about one figure in a procedure. Figures 6.7 and 6.8 illustrate some of the findings from the tables.

Tables 6.3a and 6.3b support the notion that the Gerund form is the classic choice for the title of a procedure (Farkas, 1999). It is clearly the most frequent choice of the four options mentioned by Farkas. The rubric Other includes titles such as "You need two cars," "I want my own style," "GoToPage," "How to adjust the monitor," "Mr Bios" and "Jumpers." Infinitives appear only in subtitles, but their frequency, along with that of subtitles in general, is very low in the sample.

Prerequisites

Prerequisites are conditions that must be satisfied so that the user can achieve a task. Generally, there are three main types of prerequisites, namely system states, user skills, and user knowledge. Users frequently find themselves in the position of having to assure that they are in the right system state. That is, they must know or find out what the starting position should be and what materials they need for task execution. We discuss two ways of handling theses issues under the rubric of modularity. Prerequisite skills and skills learning are especially important in tutorials. Novice users can execute some tasks only when they already know how to perform other, more fundamental tasks. We discuss fading as a technique to scaffold this learning process. Taking into account the user's prior knowledge should always be a key consideration in design. Under the rubric of mapping we discuss the role of advance organizers to build or activate relevant prior knowledge.

Copying files to a disc by means of Send To

Microsoft Windows consists of a possibility to copy a file or folder easily to a floppy.

1. Right-click on the icon of a random file or folder.

2. Choose Send To.

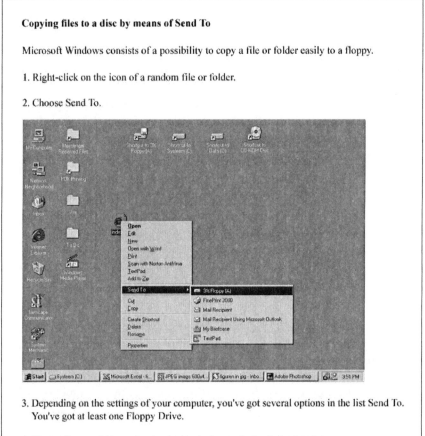

3. Depending on the settings of your computer, you've got several options in the list Send To. You've got at least one Floppy Drive.

4. Place a floppy with a appropriate amount of space in the desired station. The file will be copied to the indicated station.

Figure 6.7. An almost prototypical or 'mean' procedure: 4 actions steps, 1 figure and integrated, additional textual information (source: Cowart, 1998).

---

1 Open the Window menu and choose Mailboxes. The Mailboxes Window appears.

2 Click the Rename button beneath the listing. This calls up the Rename Dialog.

Figure 6.8. An example of action steps that also include feedback (source: LeBlanc & LeBlanc, 1995).

Table 6.3a. Meaningful goals and titles in software manuals (n=25, only the manuals in English). How is the title or subtitle phrased?

	Titles	Subtitles
Noun phrase	3	0
Gerund	14	0
Root	2	0
Infinitive	0	2
Other	5	1
None	1	22

Table 6.3b. Meaningful goals and titles in hardware manuals (n=37)

	Titles	Subtitles
Noun phrase	11	1
Gerund	16	0
Root	0	0
Infinitive	0	2
Other	10	2
None	0	32

Modularity

One of the problems that designers often face is that users want to start with intermediate chapters without having processed all the previous chapters. Facilitating such random access use of a document is no easy matter. A modular design approach aims to accommodate such uses as much as possible (Arnold, 1988; Van der Meij, 1992; Weiss, 1991). The main idea of modularity or closure is that each chapter depends on as little outside information as possible. Users should need to consult other chapters only by exception. Modularity involves two aspects of system states: starting and ending conditions of tasks, and user products.

The writer must choose a *home base* on the product's interface from which to start and end each chapter. Often this is the start screen, or window, that users see after opening a program. The main menu-bar is visible and perhaps some of the working area is displayed. The start screen is an easy point of reference from which to exit or further enter an application. In addition, it is the screen most users should become intimately familiar with. When all chapters in a manual start from home base, there is a common entry point for all users: those who come from a previous chapter, and those who just step in.

Modularity also depends on *avoiding dependency of products* across chapters. Manuals sometimes are designed around a single case, or file, the contents of which become more and more complex as users work through the chapters. Because all sorts of things can go wrong, and because users may not want to start at the beginning, such dependencies should be avoided. There are at least two ways to achieve this: Offer different sample files prepared specifically for the task(s) at hand in each chapter for users to practice on; or provide backup files users can activate when starting a new chapter which maintains the build-up of a case, or file, without the dependency.

Fading

Reference guides typically assume that users posses a minimum of basic skills for handling the program. This minimum is often the only prerequisite that users need to understand and carry out the instructions in the chapters. For tutorials the situation differs. Here users need to learn the basic skills. Skills acquisition generally requires repeated practice; most users do not learn their basic skills from completing a single exercise or instruction. This poses the problem of how to create carry-over effects. The designer must give users enough support to allow them to switch effortlessly from one procedure or chapter to another and also avoid an endless repetition of the same basic information.

One solution involves *referencing*. The user may be referred to the original, detailed instruction (e.g., "see … for a detailed description of how to …"), or to a job aid that has been created especially for that purpose. The user can consult the job aid to find out how to achieve a specific task. Job aids tend to be stripped of all extraneous information. They are bare-bone action steps (figure 6.9) that tell the user what to do. There is no information of system states or any other feedback. Just like a glossary, a section with job aids should be placed at the back of the book to facilitate accessibility.

A solution that does not involve referencing and that is more supportive of learning is fading (Carroll & Van der Meij, 1998). In fading, the support for the execution of basic skills tasks is gradually decreased (figure 6.10). From the first full instruction onward the user gets successively less complete reminders. In addition, there are other, more subtle

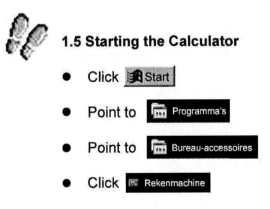

1.5 Starting the Calculator

- Click **🏁 Start**

- Point to **Programma's**

- Point to **Bureau-accessoires**

- Click **Rekenmachine**

Figure 6.9. A job aid (source: Stuur, 2000).

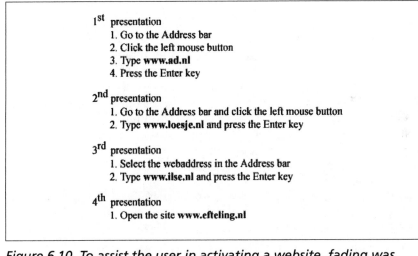

1st presentation
  1. Go to the Address bar
  2. Click the left mouse button
  3. Type **www.ad.nl**
  4. Press the Enter key

2nd presentation
  1. Go to the Address bar and click the left mouse button
  2. Type **www.loesje.nl** and press the Enter key

3rd presentation
  1. Select the webaddress in the Address bar
  2. Type **www.ilse.nl** and press the Enter key

4th presentation
  1. Open the site **www.efteling.nl**

Figure 6.10. To assist the user in activating a website, fading was used to increase learning and facilitate modularity across chapters (source: Lazonder, 2000, slightly adapted version).

changes. For example, the first action step in the 2nd presentation of figure 6.10 combines the two action steps from the 1st presentation because these form the new action step later on. In many cases one also sees a move away from a description of the physical act toward a conceptual one, as illustrated in the first action step of the 3rd presentation. Fading requires a slight mental effort from the user and thereby supports learning.

There are no fixed rules as to how many reminders users need or how fast. Only testing can reveal what works and what does not. Testing with the audience is essential to find out how quickly users can master a basic task. From piloting our own manuals we have discovered that fading can decrease mistakes and look-backs. Empirical research from Leutner (2000) indicates that fading is especially useful in the early phases of learning in which it contributes significantly to basic skills development.

Mapping

The single most important human characteristic known to affect learning is prior knowledge (Smith & Ragan, 1999). Technical documentation must therefore try to bridge the gap between the user's understanding of a task and the way in which a system forces the user to think about it. This process is called mapping. Beautiful examples of mismatched mappings in the early days of computing have been recorded by Jennings (1990). Among others, she mentions the situation of a person walking over to the Xerox workstation to "copy a diskette." The flexible diskettes also allowed people to "store a diskette on an accessible place" by tacking it to a bulletin board. Mapping problems are not bygone history. They still pervade all human–computer interactions.

Designers can use *advance organizers* such as analogies and meta-phors to handle problems with insufficient or absent prior knowledge (Ausubel, 1968). In his original study Ausubel pleaded for abstract advance organizers to provide ideational scaffolding. Later work (Mayer, 1987) suggests that concrete advance organizers are more effective in providing appropriate prerequisite knowledge. There are two main types of advance organizers: comparative and expository ones.

Comparative organizers deal with mapping problems. They serve to activate the relevant user's prior knowledge to the task at hand. In other words, they try to help the user in making the connection between real-world tasks and computer tasks. For example, a manual (not from the sample) tells users that "A HyperCard file is like a book. Each page has different words and pictures, yet together they have one purpose: to tell you a story, for example. But instead of reading pages, you read cards. A card is a screen full of information—no more, no less. And instead of turning pages, you click on buttons to get from one card to the next. The 'books' you create with HyperCard are called stacks" (Jones & Myers, 1988, p. 2).

Advance organizers can also be very useful for users who have no relevant prior knowledge. In such a case an expository advance organizer can be used to build the knowledge base. Such organizers often present the user with information about underlying concepts, or with an explanation of what an option can do. For example, a manual (not from the sample) tells users "With macro's you can execute tasks automatically in Access. Macros are often linked to buttons on forms. For example, you can create a button for generating a report of the monthly sales and a button that allows you to return to the main menu" (Bruck, 1999, p. 456).

Minimalism cautions against an uncritical use of advance organizers. Because documentation should respect the integrity of the user's activity this may mean that in some cases the presentation of information or explicit instruction must be subordinated to the continuity of the user's project-oriented activity. In addition, designers should keep in mind that the moment-to-moment goal of the user may be much less sophisticated, much shorter term than that of the designer (Carroll 1990; Mack, Lewis, & Carroll, 1990).

Recommendations for practice

Design guidelines for the Prerequisites component are:

- Chapters should be modular as much as possible.
- Use fading to increase learning when skills development is important (e.g., in tutorials).
- Advance organizers can be useful to help the user in mapping between real world tasks and computer tasks.
- Advance organizers can be useful to build prerequisite knowledge.
- Respecting the user's activity may require the designer to abstain from presenting conceptual information.

Results from the inventory

On modularity

Modularity is a difficult issue for designers. Creating modular documentation means attending to the intricate relationships that exist between tasks, task components, and underlying concepts. Modularity is also difficult to chronicle. Our difficulties in documenting this issue partly stem from having sampled only one procedure from each manual. Within a larger whole (e.g., one or more chapters) the treatment of modularity may be clearer than from a single, decontextualized procedure. For this reason we also cannot report the findings for fading, as the decline of support over procedures cannot be observed when one looks at only a single procedure. The inventory examines two issues of modularity, namely starting states and references.

One of the ways to achieve modularity comes from describing or depicting the system state right before the action section. In software manuals we found only two instances of such descriptions (table 6.4a). One of these cases is shown in figure 6.11. Indicating the starting states is more common in hardware manuals (table 6.4b). We found seven instances in which the starting state was explicitly mentioned. For example, one manual indicated that "The function Full Size works only with the factory default settings" and there were variations on "To install … you need the following …." In addition, there are two frequently occurring situations in which the absence of any indication of a starting state seems obvious in hardware manuals. One is when the user is instructed to shut off power in the first action step. Often this is a precaution, and simply by performing that act the user creates the right starting condition. The other situation involves inserting a floppy or CD-ROM into a drive and similar actions. Except for the novice, such actions are self-explanatory and often do not require a description of the system state. In all, however, the presence of information the starting states is rather meager and the results signal that the modularity issue probably needs more attention.

Full modularity or complete independence is almost impossible to realize. Indeed, such independence is not always desirable because it inevitably entails frequent repetitions. One way to make connections involves the use of references. Such references either refer to information within the same document or to another manual or online help. References within the same document may refer backwards or forward. There were three instances of referencing in the software manuals. One of these was a forward-oriented reference to elsewhere in the same document (figure 6.12), one was an outside reference to online help and one referred the user to an earlier section in the manual as well as to an outside source. There were six instances of referencing in the hardware manuals (figure 6.13). All these references directed the users to another manual. In one instance a caution was contained in the reference (figure 6.14).

To get the most out of this chapter, start a new presentation in Harvard Graphics and follow along, leaving the program running. To do this:

1. From the main Harvard Graphics window, select **File | New presentation or Ctrl-N**. The Add Slide dialog box appears.

Figure 6.11. A description of the required starting state (source: Harrison & Yu, 1992).

Table 6.4a. Prerequisites in software manuals (n=52). Which signs of modularity are used and how often are advance organizers used?

	N	%
Showing or describing the starting state	2	4
Referencing to elsewhere	3	6
Advance organizers	7	13

Table 6.4b. Prerequisites in hardware manuals (n=52).

	N	%
Showing or describing the starting state	7	13
Referencing to elsewhere	6	12
Advance organizers	1	2

 For more information about setting the tab order, see Chapter 8, "Creating and Customizing Forms."

Figure 6.12. A clearly signaled forward-oriented reference (source: Valentine, 1993).

On mapping

Mapping is intended to connect the user's understanding of a task with what the system or apparatus requires or affords. By describing what can be achieved, goal descriptions try to bridge the gap between the two. Typically such descriptions tell the user "You can use this for ..." or "You can do ... and ... with this." For example, "With Full Size you can enlarge the size of your monitor up to its maximum." Goal descriptions stay "within the system" that they describe. Thus they afford mapping, but do not actively suppot this process. Goal descriptions may need to be complemented with organizers to realize mapping.

Expository advance organizers provide users with information that explains how things work. Expository organizers, just like goal descriptions, stay "within the system." They try to build understanding and hence create prior knowledge by explaining the inner works of the program or computer.

The Gravis Analog Pro Joystick has 5
buttons, two of which are adjustable.
Buttons 3&4 will appear as Buttons
1&2 of Joystick B. See your Analog Pro
User's Guide for more information.

Figure 6.13. A reference to another document (source: Advanced Gravis Technology, 1994).

If you have only one hard disk drive and one CD-ROM Drive,
Each jumper setting is as follows:

Hard Disk Drive: MASTER
CD-ROM Drive: SLAVE

Caution

Please refer to the manual of hard disk drive for setting
"MASTER"

Figure 6.14. An outside-oriented reference cast in the form of a caution (source: Tae il Media, 1996).

Comparative advance organizers, in contrast, go "beyond the system" to help the user in bridging the gap between the system and the real world. Comparative organizers can be analogies or metaphors that are specifically tuned-in on the relevant prior knowledge that users are likely to have. These organizers are called comparative because the reader is stimulated to compare systems.

We found seven advance organizers in the software manuals (table 6.4a), all of the expository type. Figures 6.15 and 6.16 illustrate these cases. Figure 6.15 is special because the user is referred to the scientific notational system. The organizer is classified as expository because we assume that the majority of readers has no knowledge of this convention. In the hardware manuals we found only one advance organizer. This too was an expository organizer.

Unwanted states 1: Warnings

Users are prone to make mistakes, which is why a model of procedures must concern itself with unwanted states. In our Four Components model we split these unwanted states into two distinct types: warnings and problem-solving information. Warnings are given to prevent certain actions of the user or to alert users to the presence of a more or less

Big numbers are printed scientifically because the default
column size cannot display more than eight digits. This
means that the figure 1.000.000.000 is represented as 1.0E+09.
Once you expand the column width, Lotus shows the figure
in its normal reading form.

*Figure 6.15. An expository organizer that explains the 'inner works'
of a program (source: Evers, 1994).*

Downloading Fonts

PostScript LaserWriters have 35 fonts built into them. When you want to
use an outline font that's not built in, you simply use it in your
document, assigning the font from the Font menu. When a document is
being printed and the Mac discovers that the font it needs isn't in the
printer, it looks for a printer font file that it can *download*, or send, to
the printer.

*Figure 6.16. An expository organizer that explains the 'inner works'
of a program (source: Zadetto Aker, 1991).*

serious risk. Industry has always given considerable attention to the
design of warnings. Various international standards describe the directions for presenting warnings in some detail (e.g., ANSI Z535; DIN 8418;
ISO 37; ISO 3864). In addition, there is a sizeable body of research on the
design of warnings. We will go beyond a restricted view of warnings in
procedures because the design of warnings should always be part of an
overall strategy of safety measures. Before we begin, let us first consider
the broader context.

The optimal sequence for dealing with hazards is (1) remove from
the design, (2) guard against, and (3) warn (Laughery & Hammond,
1999). The first preference is to eliminate the hazard by alternative
design. When that is not possible, guarding is the second line of defense.
Its purpose is to prevent contact between the user and the hazard. Only
when guarding too is not possible, warnings come into play. Thus,
warnings are the last line of defense. They are so because "influencing
behavior is sometimes difficult and seldom foolproof" (Laughery &
Hammond, p. 5).

Warnings inform people about "what happens if." Their general
goal is to improve safety by influencing people's behavior or helping
them make an informed decision about the risk, as in the case of a
warning on a cigarette package. Noncompliance can have serious
consequences for the user. In addition, liability poses a serious financial
risk for the vendor. Designers of technical documentation must therefore

often comply with strict company rules or (international) standards. Wogalter and Leonard (1999) indicated there is a need for a standardized minimum for warnings but they caution against inflexibility. Standards and guidelines are best seen as good starting points that may need modifications now and then. One of the reasons why there is a need for flexibility is the risk of habituation. Too much similarity across warnings may eventually lead to a familiarization which can result in a dramatic decrease of warning effectiveness.

The design guidelines in the literature generally are derived from models of communication and models of human information processing (Wogalter, Dejoy, & Laughery, 1999a). The communication model addresses the concepts of source or sender, channel and user or receiver. Information processing theories decompose the receiver's processing of a warning into stages of attention, comprehension, motivation, and behavior. Because the aspects of source and channel are fixed by our choice for paper documentation they are not detailed here. We concentrate on information processing. More specifically, we use the "see-think-use" model for grounding the design guidelines for warnings.

The See–Think–Use Model

The main stages in processing warnings have been nicely captured in the "see-think-use" model that we will present here. [For a more detailed model, as well as extensive references, the reader is referred to Wogalter Dejoy, & Laughery, 1999b.] Just as its name implies, the model assumes that there are three main stages. Users must first perceive (see) the message, then they must understand it (think), and finally they must act accordingly (use). A main goal of warnings is to induce compliance. Mere perception and understandability of a message is not enough; users must also be motivated to act safely. Instead of adapting the see-think-use model, we accommodate to the role of motivation by including this factor in the think stage (figure 6.17).

The see–think–use model suggests a linearity that is valid only in certain cases. One of the problems with linearity relates to the user's prior knowledge. For example, a knowledgeable user may ignore the major part of a warning because its mere presence already cues that person to act safely. The opposite may also be true. A knowledgeable person may see the warning, believe it to be irrelevant and ignore its message. Even so, the model is useful for the design of warnings in a proactive as well as reactive sense. As a proactive means, it attends the designer to critical factors that should be taken into account in the design of warnings. As a reactive means, the model can serve as a diagnostic tool for discovering why some warnings are less effective than others.

1. Seeing a warning
   • Capturing the user's attention
   • Maintaining the user's attention
 	
2. Thinking about a warning
   • Comprehending the warning
   • Remembering the warning
   • Complying with the warning
 	
3. Using a warning
   • Acting upon the warning

Figure 6.17. An information-processing model of warnings.

Stage 1: Seeing a warning

First of all, the user must perceive or see the warning. The warning must be attended to. This typically involves two stages, namely drawing the user's attention and maintaining it. To *capture the user's attention* the warning must be conspicuous. It must stand out from other stimuli. The key to capturing the user's attention is salience. To achieve saliency designers can use contrast, highlighting, size, signal words (e.g., caution or danger), pictures, and location, among others.

In technical documentation the role of location is special because existing practices tend to stand squarely on the general rule that advocates an approach in which a warning should be conspicuous and given timely and close to the hazard. Company policies may dictate that the warnings should be presented in a separate section up front in the book. This practice seems to conflict with the view that the noticeability of a warning increases when it is integrated in a procedure. Research has yet to give a clear answer as to which approach works best. Studies have yielded equivocal results on the value of selecting for an integrated or separated approach. The tentative conclusion is that an integrated approach is preferable when the user is not familiar with the task and the risk is high (Wogalter & Leonard, 1999).

Warnings are ineffective if they draw the user's attention only for a very short moment. Users should not opt out immediately afterwards but keep on attending to the warning. This second phase in seeing the warning is called maintenance. *Maintaining the user's attention* means that the warning must hold the attention long enough for the user to read (parts of) the text and examine the picture. Two critical factors in this regard are legibility and readability. Legibility involves the physical aspect of perceiving and distinguishing separate features of textual and pictorial elements. This factor is relevant for software and hardware manuals only in very special situations (e.g., bad lighting conditions). Readability refers to the conceptual aspect of information processing. It refers to ease of reading and its corollary of content or meaning.

Stage 2: Thinking about a warning

Warnings must inform the user of various issues of hazard control. That is, a fully informed user must comprehend the hazard, know how to avoid it, and know the potential consequences of unsafe behavior. In addition to filling a knowledge gap, users may also need warnings as a reminder or cue to prevent them from forgetting to act safely at the critical moment. In short, warnings must be *comprehensible and memorable*. There is extensive research, which suggests that the combined strength of text and picture can best achieve this.

A critical problem with only textual information is the vagueness of terms. Industry has attempted to handle this problem by suggesting the use of the standardized signal words. In the US, these are the words caution, warning and danger, which connote low to high levels of hazard (ANSI, 1998). However, because most people do not know the formal definitions associated with these words, they mainly alert the user and produce an overall impression of hazard (Leonard, Otani, & Wogalter, 1999).

A critical problem with only pictorial information is that pictures are not as easily understood as is sometimes assumed. This raises the question what level of comprehensibility is acceptable. The American National Standards Institute (ANSI, 1998) requires an 85% success criterion for pictures with no more than 5% critical confusion. Any score below that level requires an accompanying text. The European Organization of International Standards (ISO 3461-1, 1988) requires a 67% success criterion for just the picture.

Many people who have noticed and read a warning still fail to comply. For example, Friedmann (1988) reported a study in which 88% subjects noticed a warning, 46% read it, and only 27% complied with the warning. Acting safely could easily be done and therefore, the findings clearly signal that motivation was at stake. This and other studies indicate that warnings should induce *compliance*. A warning should be persuasive to ensure the correct attitudes and beliefs. The factors that are in play here concern the user's assessment of the likelihood and severity of the hazard, and the effort needed to act safely. Value-expectancy theory captures these factors in a model, which holds that people estimate the seriousness of a risk, evaluate the costs and benefits of various actions, and then select a course that maximizes the outcome.

An important rhetorical factor in this respect is source credibility. One situation in which people are not easy to convince is in medical leaflets. For example, the following warning: "Note. This medicine causes drowsiness. Driving a car and the use of machinery may be dangerous" often does not evoke the reaction "I won't drive." Rather it stimulates people to think "I must go back to my physician, this stuff is clearly not suitable for me." (Karel van der Waarde, personal communication).

Stage 3: Using a warning

The ultimate goal of a warning is that the user acts as safely as possible. The user should avoid certain actions, or take precautions to minimize the risk. To achieve this, a warning must inform the user of the most appropriate actions. As indicated earlier, the location of the warning is probably a critical factor. When the warning involves an unfamiliar task, a position right before or within the directions for use, is a viable option (e.g., Frantz, 1994; Friedmann, 1988; Strawbridge, 1986; Venema, 1989). Users who read such a warning are more likely to act safely because the information is immediately useful. It is "just in time" information. The opinions vary as to whether the warning should stand out from the other instructions when it is integrated in the workflow (e.g., action steps).

Research indicates that some users read only the first sentence or the first few words of a warning and then move on to the next step in a procedure (e.g., Franz, 1994; Friedmann, 1988). Obviously this can seriously jeopardize the effectiveness of the warning. In this respect there is an interesting study on the positioning of the instruction with regards to the description of the risk. This study challenges the validity of the conventional order of presenting the risk description before the instruction.

Participants in the study by Maes, Maes, Van der Meulen and Verbunt (1998) were asked to rate risk perception, tendency to comply and naturalness of warnings with Instructions before Risk descriptions (e.g., "never pull out jammed papers from the printer by hand, the printer may get damaged") and warnings with Risk descriptions before Instructions ("the printer may get damaged, never pull out jammed papers from the printer by hand"). The results clearly favored a presentation in which the instruction preceded the description of the risk. Analyses of the data further revealed that the effect on naturalness of the Instruction–Risk order of presentation was stronger for risks involving a personal injury than for product damages. The authors suggest that this order of presentation is more effective because it emphasizes the instructive nature of warnings.

Dixon's (1982, 1987) basic research on the "use-order principle" also supports this stance. His experiments show that instructions are read faster when the action information precedes conditional information. This is true regardless of whether these sentences refer to the antecedent or the consequence of the action. This being the case, Dixon (1982) concluded one of his articles by expressing the view that "where time is important 'break glass in case of fire' may be preferable to 'in case of fire, break glass'" (p. 83).

Recommendations for practice

Design guidelines for the Unwanted states—Warnings component are:

- Warnings should be signaled with text and a picture.
- Warnings should include a signal word.
- Warnings should be brief.
- Warnings should be easy to read.
- Warnings should be persuasive.
- Warnings should describe the hazard.
- Warnings should describe the consequences of noncompliance.
- Warnings should instruct the user what to do or how to avoid the hazard.
- Warnings should present the instructions immediately after the signal word.
- Positioning a warning right before or in the action steps is preferable especially when the task is unfamiliar and the risk is high.

Results from the inventory

On the See-Think-Use model

Tables 6.5a and 6.5b present the findings on what are probably the minimum acceptable standards for presenting warnings (see Wogalter, 1994), namely:

- The warning stands out to facilitate perception.
- A signal word is included to help users recognize the information as a warning and to indicate the level of the hazard.
- A description of the hazard specifies the risk involved.
- Information about the consequences helps users assess the effects of non-compliance.
- Instructions tell the user what (not) to do.

The first thing to note about the data is that the presence of warnings in the sample seems a bit low. Only 20 of the 104 manuals included a warning. This may partly come from our coding system. There were two types of situations in which we felt the information might have been called a warning, but lack of signaling or ambiguity of the information content prevented these from being coded as a warning. One of these situations is when users are instructed to switch off an apparatus. The other situation is when users are informed about measures that can prevent relatively minor inconveniences.

Table 6.5a. The main characteristics of warnings in software manuals (n=8).

	N	**%**
The warning stands out on the page	5	62.5
There is a signal word	4	50
The hazard is described	2	25
The consequences of non-compliance are described	5	62.5
The instruction of what to do or to avoid	8	100

Table 6.5b. The main characteristics of warnings in hardware manuals (n=12).

	N	**%**
The warning stands out on the page	8	66.6
There is a signal word	8	66.6
The hazard is described	2	16.6
The consequences of non-compliance are described	6	50
The instruction of what to do or to avoid	10	83.3

Quite a few hardware manuals instructed their users at one time or another in the actions to switch off their apparatus. In 10 manuals this instruction was given in the first action step; in two manuals this instruction appeared somewhere in between these steps. None of these manuals indicated that this action was, as we presume, a precaution needed to prevent shock or product damage (figure 6.18). By *not*-signaling and *not*-describing the risks involved, the designers presented the information as just another action step. This may increase the chances that users will heed the advice and act as indicated. However, there is a risk of loss of source credibility when users realize that the action is a precaution and not a regular action step. In general, designers should not mix different information types into a single mold or format.

The other instance in which we hesitated about calling information a warning concerned situations in which users were informed about measures they should take to prevent minor problems such as losing data by not saving a file. We found two instances in which users were advised to take precautions. We coded these messages as a warning only when their presentation format differed from other information types in the procedure.

About 65% of the warnings are easy to perceive thanks to their presentation format (e.g., in italics, another font), a picture or a signal word such as "Caution" or "Warning" (figure 6.19). Handling apparatuses may involve serious risks and for this reason the percentage of unsignaled warnings in hardware manuals is deemed too high. Examples of these unsignaled warnings are "Confirm the line voltage designated on the rear panel of the monitor" and "Discharge any static electricity from your body by touching any metal surface." In both cases this information was presented as regular text in the introductory paragraph of a procedure.

Not all the signaled warnings are easy to perceive. Figure 6.20 shows a signaled warning that is not very noticeable. Partly this is due to the small print. We have coded the information as a warning because the designer has given the information a distinct presentation format. Considering our conservative coding system, we believe it is fair to say that perhaps as much as about 50% of the warnings in the hardware

---

1. Turn off the Computer.
2. Remove the blanking plate available for disk drives.

Figure 6.18. The user is instructed to turn of the power in the first action step. In this way compliance may be increased but there is a risk of loss of source credibility when users recognize this as a precaution and not a regular action step (source: Tae il Media, 1996).

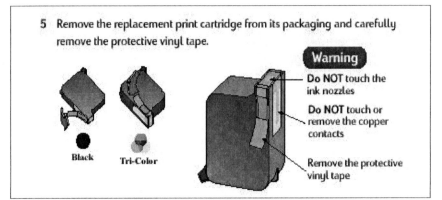

Figure 6.19. A signaled warning within an action step (source: Hewlett Packard, 1996).

manuals are not adequately marked. Their users are likely to risk damaging products or worse. In addition, designers or makers run a liability risk.

The majority (60%) of the warnings were placed between the action steps rather than before or after these steps. For software manuals the percentage of these "on-the-spot" warnings is considerably lower than for hardware manuals (respectively 37.5% and 69.2%). This is comprehensible if one considers the different nature of these warnings. For example, the two warnings that came after the action steps both stemmed from software manuals and both informed users of a risk of losing data. The warning displayed in figure 6.21 illustrates such a case. This warning formed the end of a procedure in which users were instructed to rebuild their desktop as a protection against viruses. Oddly enough this argument was not included in the warning.

The different percentages for the factors risk description and specification of the consequence may have to with our coding. We defined the risk factor as a state and coded the consequences as a type of (system) action. Even so, these two features of a warning were sometimes hard to distinguish from each other. All warnings stipulated what users had to do or not do, to avoid running a risk.

Install the CD-ROM drive into the unit and secure with mounting scres (recommended standard M3x5)

   ➤ ◆ Depending on your PC model, you need to attach guide rails to the drive, so that the drive is held firmly in place.
    ◆ If you use screws that are longer than the recommended standard, you may damage the drive

---

Figure 6.20. The signaled warning is not easy to perceive. Note that the warning is not clearly recognizable as such in the absence of any standardized symbol or signal word (source: Samsung, 1996).

Rebuilding your desktop file erases any comments in your Get Info dialog box. Rebuilding your desktop file improves your Macintosh Finder's performance, and is a good habit to get into. Rebuild your desktop every few weeks.

Figure 6.21. A warning in which users are advised to rebuild their desktop regularly to improve performance. Earlier on in the procedure the user is told that such an action helps protect against viruses (source: Hanson, 1993).

Actions & Reactions

Constructivist models of instruction aim to improve learning by creating situations in which people actively construct their own understanding. In the field of technical documentation the minimalist approach exemplifies such a constructivist view (Carroll, 1998). One of its premises is that the effectiveness of a manual depends on finding the proper balance between support and let go. Users benefit from guided practice. They need direct instructions to act as well as guidance to support their explorations. We briefly discuss this important principle. An equally important principle for the design of the component Action & Reaction is the interaction between user input and feedback. To illustrate the intricate patterns of what users must do (or think) and how the system reacts, we discuss how one can optimize the design of screen captures to the roles at hand. We then zoom in on lower level aspects of action steps. That is, we discuss the desirability of numbering action steps that are to be executed in consecutive order and we present some views on how to write these steps.

Direct instructions and guided discovery

The designer often needs to find the right balance between direct instructions to act and indirect ones such as invitations to explore. Balancing is needed to meet the propensity of users to explore. The designer has to satisfy this tendency up to a certain level, or else run the risk of losing the interest of the audience. Another compelling reason for balancing direct instructions and guided discovery is that such a mixture affords the development of strategic knowledge. Under such conditions self-directed learning can lead to deep-seated knowledge. Users learn not just "how to do it" but also "how it works".

The shift toward supporting basic skills development through meaningful exercises in combination with advanced skills development through guided exploration is described in a variety of approaches (e.g., Alpert, Singley, & Carroll, 1995; Bhavnani, Reif, & John, 2001; Bhavnani & John, 1997, 2000; Lazonder, 2001a, 2001b; Mirel, 1998). An important problem in designing for guided discovery learning lies in creating the right support or context. Users must have enough knowledge to form appropriate goals, pursue relevant activities, and make the right inferences. In addition, there should be a guarantee of safe progress (e.g., by scaling down the initial problem space in which users will be acting) and the explorations must be motivating as well as tractable.

An example from a relatively simple domain comes from our work on word processing. The "on your own section" in figure 6.22 does not merely ask students to explore. It cues them to consider what other goals they might want to pursue in relation to searching a text. The suggested goals for exploration are related conceptually and procedurally to the operations practiced in that context. The user has just been thinking about forward search, and may therefore be prepared to discover backwards search. The exploration is also quite tractable: In this case, searching backwards requires only two slightly different actions (i.e., positioning the cursor at another place, and selecting another option from the Search menu). A similar choice of basic skills training with simple tasks followed by strategies that rely on these skills can be found with Bhavnani, Reif, and John (2001) and Wiedenbeck, Zavala, and Nawyn (2000).

The same design principle, in a slightly different guise, can be found in the complex domain of Smalltalk programming. Carroll and his colleagues created a Guru (an expert programmer) to help novices exploit more fully the capacity of existing (sub)programs in Smalltalk (Alpert et al., 1995). Users could call up this Guru at the end of a tutorial project. Among other suggestions the Guru could advise that the user makes a more efficient use of the inheritance mechanism in Smalltalk and convey the insight that this expert strategy is easier to maintain. The Guru was available only at the end of a project because the authors believed that only then were users most receptive to the commentary and in the best position to appreciate the insights.

Research indicates that a proper preparing and cueing strongly encourages learners to explore. For example, one study found that such cues induced 81% more exploratory episodes than occurred with a

Searching a text	You can position the cursor quickly to a word or part of a sentence by searching for this text.
	1. Position the cursor at the beginning of the file
	2. Go to the member and press twice on the → key
	3. Choose the command FORWARD and press the ENTER key
	WordPerfect asks what you want to search for. Check to see if the prompt ⇨ Search: is on your screen.
	4. Type any word(s) from the text
	5. Press the F2 key
On your own: Searching text	The command NEXT and PREVIOUS enable you to find out if the word you have been searching can also be found elsewhere in the text. You can find these commands under the SEARCH option. Try them and see.

Figure 6.22. When users are invited to explore on their own they should have adequate prior knowledge and skills. (source: Carroll & Van der Meij, 1998, slightly adapted version).

control manual (Carroll, Mack, Lewis, Grischkowsky, & Robertson, 1985). Another study found that 74% of all invitations to explore in a text-processing manual actually led to explorations by the users. In contrast, the invitations to explore in a control (i.e., conventional) manual were significantly less tempting with an average compliance of 41% (Van der Meij, 1994). Research also shows that users acquire more strategic knowledge with a properly balanced approach (Bhavnani, Reif, & John, 2001; Thomas & Foster, 2001; Wiedenbeck, Zila, & McConnell, 1995).

The interaction between user actions and feedback

Feedback is vital in all systems. Procedures are therefore filled with intricate action–reaction patterns. There are switches back and forth between user input on the one hand and system reactions and other kinds of feedback on the other. We present the research on screen captures to illustrate the nature of these action–reaction patterns. Rather than treating feedback as a separate object of study, the research emphasizes the need to consider the interaction between what the user must do (or think) and how to design for that. More specifically, the research indicates that screen captures can serve four main roles and that their designs can be accommodated accordingly.

The work on screen captures fits within a long tradition of research on the roles of instructive pictures. Important theories in this domain are dual coding and cognitive load. According to dual-coding theory, people process verbal information differently than they do visual information. Learning can be optimized when the two modalities are combined in such a way that they strengthen each other. By using the capacity of the two systems, people can learn more than from using just one system. In addition, their simultaneous processing improves their connectivity and leads to better outcomes such as the formation of a mental model (Mayer, 1999; Paivio, 1990).

Cognitive load theory draws the attention to the limits of people's working memory. In complex situations the designer must pay special attention to two factors that may negatively affect working memory, namely, redundancy and split attention. Redundancy effects occur when users are confronted with two or more presentations of the same information. Screen captures repeat what the user can see on the actual screen and according to cognitive load theory they may thus hinder rather than help the user. Split attention effects occur when people must attend to different sources of information simultaneously. When these distinct sources must be integrated, cognitive load may become too high and learning is impeded (Chandler & Sweller, 1991; Sweller, 1994; Sweller & Chandler, 1994; Sweller, Van Merrienboer, & Paas, 1998).

In the minimalist approach to documentation the presence of split attention problems has long been recognized as an important obstacle for first-time users. Indeed, the observation of the problem even prompted the research on screen captures in technical documentation. Early studies found that, under certain conditions, these pictures could alleviate split attention problems and related obstacles in human–computer interaction (Van der Meij, 1996a; Sweller & Chandler, 1994).

Later work further examined the potential contributions of screen captures and yielded a taxonomy that describes four roles and four design dimensions (Van der Meij & Gellevij, 1998). The four roles are mental model development, switching attention, verifying screen states, and identifying and locating window elements and objects. These roles accord with Mayer's SOI-model (1999) in which pictures, along with text, are used to help users §elect, Organize, and Integrate information in working memory and build a mental model or schema. The four design dimensions mentioned in the taxonomy are: coverage, position, size, and cueing. The following sections describe which designs best accommodate these main roles of screen captures.

Screen captures can support *mental model development* when they assist the user in acquiring the look and feel of a program and when they help the user understand its underlying structure. To facilitate this development the screen captures should depict *system topology* and *component behavior* (Mayer & Gallini, 1990). System topology simply means that the screen capture shows the important elements of a screen and positions these in a wider context. The elements can be windows, parts of windows, and icons; the context can be a toolbar, the home page of a program or the computer desktop (figure 6.23).

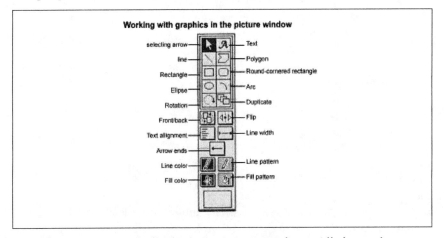

Figure 6.23. A picture displaying system topology. All the main components are shown in context and labeled (source: Heid, 1992).

According to Mayer and Gallini (1990) the best picture for displaying component behavior is a figure that shows two features, namely, the labeled key elements and the flow. The labels denominate the elements. Flow is represented by describing the actions on these elements and the consequences of these changes on the system state. It is important to note that Mayer and Gallini examined people who were reading for understanding. The participants in their study were not asked to perform a behavioral task. In manuals the link with instructions is often more direct, which may reduce the need to integrate the two features. Figure 6.24 shows a screen capture that does show an integration.

Screen captures can help reduce what is known as the "nose-in-the-book" phenomenon. By their very nature screen captures invite users to *switch attention*. The pictures clearly tell the user when it is important to look up from the manual to the screen. The need to switch attention automatically brings along the difficulties of split attention. Screen captures can reduce the negative effects when their design integrates text and picture. Figure 6.25 shows an integrated design in which instruction and screen capture are displayed as an entity. Notice that the design avoids redundancy. What is depicted is not also presented textually. The absence of a wider context also contributes to switching because the pictures are not self-contained; they do not show the key elements in context.

Screen captures can also help users in *verifying screen states*. When users consult screen captures and discover that they are still on the right track the pictures serve as positive feedback, which reinforces motivation. Especially for the novice user this may be important to allay initial anxiety. Apart from checking progress, users can also use screen captures to verify whether the program has processed their input correctly. The special advantage of pictures compared to text is that the user needs merely to compare the computer screen with the depicted screen. The comparison is direct; there is no need to translate a description. The main design consideration for the verifying role of pictures is legibility. The picture's size must be large enough to afford an easy check of information. Occasionally cueing may be desirable to attend the user to the section that needs to be verified.

Nearly all screen captures give information with which users can *identify and locate window elements and objects*. This is no luxury considering the complexity of most user interfaces. Among the multitude of menu options, windows, icons, and symbols the user has no easy task in picking the element or object that is needed. Making the right choice hinges on two factors: knowing which element to look for and knowing where it is situated. Screen captures can help with this identification process as well as with finding the location on the screen. Although the

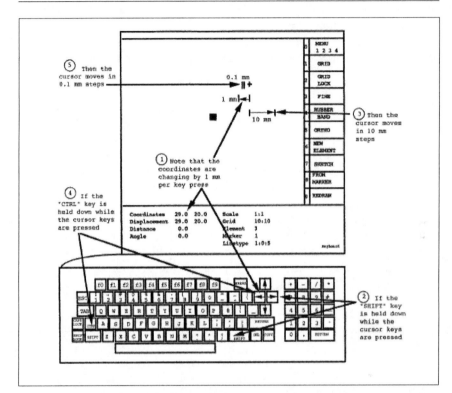

Figure 6.24. An example of a picture displaying component
behaviour. The main objects are shown and labeled. In addition, the
action sequence describes flow (source: Sweller & Chandler, 1994).

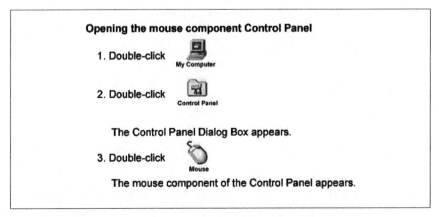

Figure 6.25. A screen capture designed specifically to support
switching attention.

main design consideration for each of these processes is slightly different, the two are often combined. That is, window elements are shown in a wider context to facilitate object location and cueing is used to facilitate identification. Just as in balloon help, callouts are sometimes used to add an explanation.

Research generally shows that users benefit from the presence of screen captures. For example, Van der Meij (1996a) reported that the users of a manual with screen captures completed training significantly faster than did users of a textual manual (see also Martin-Michiellot & Mendelsohn, 2000). The study even reports an effect size of more than one standard deviation. Similarly, Sweller and Chandler (1994) found that screen captures reduced training time and improved learning outcomes.

Recently, research has been split into two tracks: design-based studies and role-based ones. Design-based studies depart from a genre approach. That is, the design of the screen captures is standardized throughout the manual. Such an approach sacrifices the possibilities of optimizing the designs to the roles at hand for a less time consuming, more uniform approach. All design-based studies show that full screen captures are preferable to partial ones. For example, Gellevij, Van der Meij, De Jong and Pieters (1999) found that the presence of full screen captures improved learning more than did partial screen captures. And in a comparison of three visual manuals, Van der Meij (2000) found the best effects for a manual in which the coupling of instruction and screen captures followed a left to right reading order and in which all screen captures were full screen shots. Users who worked with the Instruction–Full Screen manual completed training at least 25% faster and showed at least 60% better retention afterwards.

A recent role-based study shows that optimized designs for screen captures can have a strong effect on the user (Gellevij, Van der Meij, De Jong, & Pieters, 2002). Just as in other studies the presence of screen captures significantly reduced training time. The pictures also helped create a stronger mental model, improved the identification of window elements and objects, and speeded up locating these. The magnitude of these effects were considerable with effect sizes of between 0.74 and 0.84 standard deviation. No positive or negative effects on cognitive load were found. The presence of screen captures neither reduced the load on memory nor did it increase this load to an unmanageable proportion. None of the studies to date has found any positive effects of screen captures on user motivation, however. Apparently the pictures do not play a vital role in allaying anxiety.

Design-based studies and role-based studies each contribute relevant insights for technical documentation. Further research on the impact of design variations can contribute directly to improvements of the genre approach. Role-based studies give a more detailed view of the fundamental processes that underlie action–reaction patterns. An extension of these studies with user observations should yield additional insights for optimizing user support.

Signaling actions

Actions should be signaled. Perhaps the most critical reason for this design guideline is ease of access. In tutorials as well as reference guides re-reading of instructions is likely to occur regularly because users generally remember what they have done only after repeated exposure and training. Each time a user wants to perform a task for which the actions have not yet been automated the user may consult the manual again. For this reason, actions should not be written as if for one-time consumption (e.g., in narrative format).

There are many ways in which a set of actions can be presented in a format that make the actions stand out from other information presented on a page. Three distinctly different formats are listed steps, decision tables, and flow charts. The choice of format may depend on what has to be documented. For example, a flowchart format is a good choice when a task has to be executed only once (e.g., with installation of a program) and involves a complex route with lots of branching.

Actions that must be executed in consecutive order tend to prevail in technical documentation. This should lead to a dominance of listed steps. The basic listed step presents a series of numbered actions steps. A variation occurs when these action steps are mixed with feedback such as information about the screen state. Other variations may be the use of a Play Script format or a two-column design of an action–response table.

There is unanimous agreement on the desirability of numbering the action steps in listed steps (e.g., Feinberg, 1989; Krull, 1994; Price & Korman, 1993). The following reasons are given for this preference: (a) like other forms of signaling, numbering increases the salience of, and reinforces the existing tendency to search for, task-oriented information, (b) numbers help users get a better sense of the sequential organization of a task, and (c) numbers help users find their way back if they look away from the manual in order to perform a step; they may reduce mistakes of skipping a step or doing the same step twice. Numbering can also help the designer in spotting a sequence that is becoming too long (Price & Korman, 1993).

To our knowledge only two experimental studies have directly addressed the issue of numbering (Frase, 1981; Lorch & Chen, 1986). Both studies favor the presentation of numbers. In an interesting study on the design of educational materials for low-literate parents, the authors also discovered a preference for a clearcut sequencing method. Testing revealed that most parents preferred a sequencing that progressed left-to-right and included numbers (Floreak, 1989). The positive findings for numbered lists fit within the broader view that users benefit from meaningful segmentation (Fleming & Levie, 1993; Hartley, 1994).

Directing the user through the action steps

In "Documentation for a Physical World" Krull (1994) gave a vivid description of how predominantly verbally oriented documentation may fail to support users in performing physical tasks. Can these manuals be written more effectively, he asks . They can. By following guidelines that support an engineering approach to document production. Part of the engineering process is transforming basic scientific insights and principles into useful guidelines and practices.

Drawing on a model of the stages that people go through in learning a physical task, Krull (1994) suggested that documentation should emphasize physical concepts, spatial orientation of objects and users, and involve the senses of vision, hearing, and touch. More specifically, Krull gives the following guidelines for critical issues in the design of action steps: be brief, make the user the agent of actions by using the active voice, put only one step in each sentence, make complex identification of physical objects into a step (e.g., "Search the openings on the bottom of the monitor"), break complex movements into several steps, be conscious of the difference between how an action looks and how it feels, and put steps in sequential order.

Krull (1994) is selective in the guidelines he described because of space constraints. A comprehensive treatment on creating action steps is also beyond the scope of this chapter. We discuss two issues: the use of facilitating modifiers, and the presentation of keys or screen elements in action steps.

The basic action step

In its simplest form an action step consists of a combination of verb and noun. The verb tells the user what to do. The noun indicates the object involved. The basic action step thus tells the user how to act upon an object. The user should "Press F7," "Click the Enter key," "Select Install," "Lift the handset," "Remove the Out tray," "Switch off the computer."

Facilitating modifiers

Very often users are likely to benefit from a facilitating modifier that elaborates the basic action step. According to Farkas (1999) most facilitating modifiers are locator phrases that direct the user to the location of the object. Examples of action steps with such modifiers are "Choose Download Fonts from the file menu," "In the File Manager, select Up," and "Unplug your computer from the AC outlet." The use-order principle of Dixon (1982, 1987) suggested that the preferred order is to present the action step before the modifier. That is, it may be more effective to write "Choose Copy from the Edit Menu" rather than "From the Edit menu, choose Copy."

The presentation of keys or screen elements in action steps

In software manuals the designer can choose between representing keyboard keys and elements from the screen verbally or visually. The choice is likely to depend partly on the object involved and partly on practical considerations. The general guideline should be isomorphism. When there is similarity in form, the user does not have to make a mental transformation of the object. In addition, a visual representation makes an object stand out more clearly and makes the method more elegant (Brockmann, 1990). The designer may have to compromise, however, due to practical considerations such as the difficulty to accommodate to different layouts of keys and keyboards and the near impossibility of accommodating to last minute changes of the user interface.

Menu options are least likely to be affected by a non-isomorphic representation because they are mainly verbal descriptions anyway. When keyboard keys are to be described rather than depicted, the designer should reduce ambiguity by adding the word "key." In this way mistakes such as typing the words rather than pressing the function key can be avoided. Descriptions of screen symbols and icons should be avoided as much as possible.

Recommendations for practice

Design guidelines for the Actions & Reactions component are:

- Balance the presence of direct instructions and invitations to explore.
- Prepare the user well before inviting explorations.
- Documentation should display action–reaction patterns.

- To support switching attention screen captures should form an integral part with the instructions.
- To support verification screen captures should be legible.
- To support identification and location of screen objects screen captures should present these objects in a wider context and cue these when necessary.
- A right-to-left sequence is best for presenting the combination of instructions and screen captures.
- Present predominantly full screen captures.
- Signal action steps so that they stand out from the other information on the page.
- Number action steps that must be executed in order.
- Follow the use–order principle in action steps with a facilitating modifier.
- Represent keyboard keys and screen objects visually where possible.
- When keys must be described present the word key at the end.

Results from the inventory

On direct instructions and guided discovery

The manuals offered very limited support for guided explorations. In fact, we found only two cases in which users received a more or less open instruction (figures 6.26 and 6.27). The context for exploration in these cases did not guarantee safe progress and did not assure that user have the necessary prior knowledge and skill. None of the 104 manuals presented an "on your own" section at the end of a procedure.

On the interaction between user actions and feedback

The SOI-model of Mayer (1999) suggested that two types of pictures are especially important for the development of a mental model: pictures of system topology and pictures of component behavior. In general, both types of pictures call for more than a pure display of the object. System topology requires context. The picture should show the relevant object(s) in a broader context. Component behavior requires labeling and a description or depiction of flow. The main components of a picture must be labeled and changes in the system state must be depicted or described.

For the inventory we found it useful to use a related but slightly different categorization of picture type. That is, we distinguished among labeled pictures, flow pictures, and label + flow pictures. The labels in a

2. Having a complete backup set available, you should regularly "clean out" your hard disk, deleting (or archiving to floppy disk or tape and then deleting) old, unwanted data files and applications. Opening Explorer from the Programs menu is an excellent way to do this; you can hunt through your entire folder / directory structure to delete files and even entire directories. This maximizes free space on your disk, which can improve your disk's performance and also offers the swapfile more room to provide more multitasking capability.

Figure 6.26. An example of guided discovery (source: Tiley, 1995).

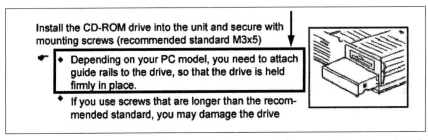

Figure 6.27. An invitation for guided discovery? (source: Samsung, 1996).

labeled figure should help users identify and locate the main elements in a picture (figure 6.28). Flow should help users see how things change from a starting state to an intermediate or ending state. In software manuals this typically means that two or more screen captures must be shown. In addition, these pictures should be shown in their consecutive order (figure 6.29).

Tables 6.6a and 6.6b show the types of pictures presented in the manuals. The tables include a flat picture category for pictures for which the presence of a label or a display of flow could have been useful (figure 6.30). In software manuals the most striking finding for picture type used is the overwhelming presence of flat pictures. These pictures are all screen captures of which the majority is positioned within the action steps. The absence of any labeling or flow probably means a missed opportunity. In hardware manuals labeled pictures are slightly more common. Figures 6.31 and 6.32 show two pictures with labels and flow. These pictures show the immense variations that can be found among these types of pictures in hardware manuals.

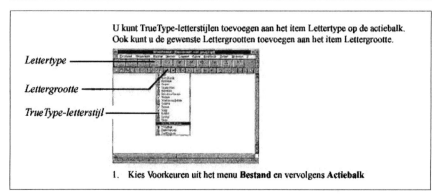

Figure 6.28. A labeled screen capture (source: WordPerfect, 1993).

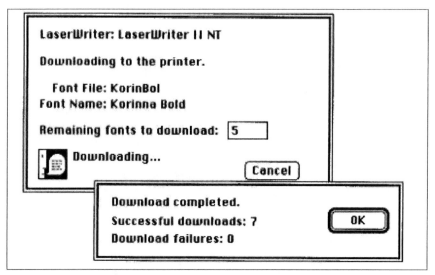

Figure 6.29. Two or more screen captures can be used to depict flow (source: Zardetto Aker, 1991).

On signaling actions

According to one of the design guidelines, instructions for use or action steps should stand out from the other information on the page to increase accessibility. The tables 6.7a and 6.7b show that the majority of designers signal the action steps. About 90% of the instructions in both types of manuals are presented in a distinct way that mark this information as a special type. Indented, numbered lists of action steps dominate with an 83%-score for software manuals and a 75%-score for hardware manuals (figures 6.33 to 6.36).

Table 6.6a. What types of pictures are presented in software manuals (n=30)?

	N	%
Labeled picture	1	3.3
Flow picture	4	13.3
Labeled + flow picture	1	3.3
Flat picture	24	80

Table 6.6b. What types of pictures are presented in hardware manuals (n=25)?

	N	%
Labeled picture	9	36
Flow picture	1	4
Labeled + flow picture	3	12
Flat picture	12	48

1. Before you connect the cables, make sure that the monitor and the system unit power switches are **off.**

2. Plug one end of the 15-pin signal cable to the monitor and the other end to the video signal connector at the rear of the system. Tighten the two screws on the cable connector.

3. Connect the power cable.

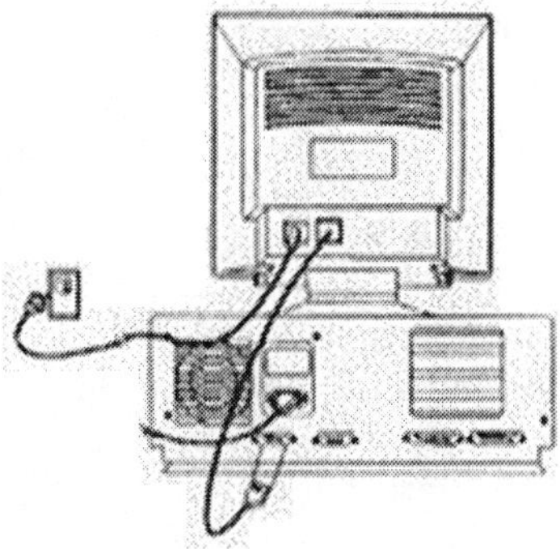

Figure 6.30. Context matters. This is especially true for 'flat' pictures which otherwise seem unrelated to the instructions (source: Acer, 1996).

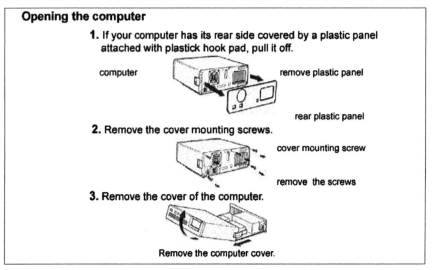

Opening the computer

1. If your computer has its rear side covered by a plastic panel attached with plastick hook pad, pull it off.

computer   remove plastic panel

rear plastic panel

2. Remove the cover mounting screws.

cover mounting screw

remove the screws

3. Remove the cover of the computer.

Remove the computer cover.

Figure 6.31. A picture combining labeling and flow (source: Sony, 1994).

The majority of the action steps are presented as numbered lists. This accords with the general rule that consecutive actions should be presented in a listed form. Seven of the bulleted lists also involved consecutive actions, however. These presentations would improve with numbering. The other bulleted lists were either a single action step, or an enumeration.

On the basic action step

In its simplest form an instruction asks the user to act on an object (figure 6.33). Verb–noun combinations such as "Click OK," and "Press the ON / OFF button" therefore always constitute the core part of an action step. Considering their vital role, designers may want to signal the core part of an action step to highlight this information for the user. In other words, they could signal the verb, the noun, or the verb-noun combination to distinguish that information from an explanation or feedback.

The inventory showed that designers hardly ever signaled the verb or the verb-noun combination. Verbs were signaled in 15% of the manuals and verb-noun combinations in 5.7% of the manuals. In contrast, signaling of the noun, the key object(s), was quite common in all the analyzed action steps (including elaborated versions). Tables 6.8a and 6.8b show the preference for type of signaling. For both software and hardware manuals, designers clearly prefer to signal the existing font. The preferred choice here is to capitalize the first letter(s) of the object (e.g., "Document," "Enter," "Computer," "Sound Card") with a 49% preference-score for software manuals and a 38% preference-score for hardware manuals. This preference for capitalization probably partly

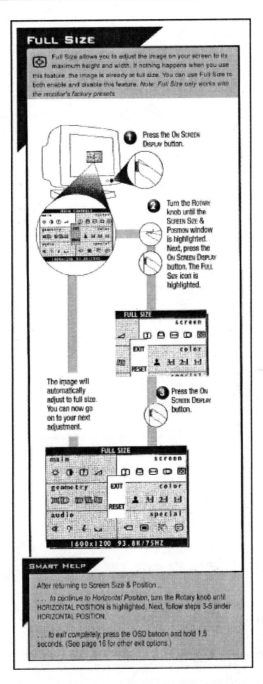

Figure 6.32. A picture combining labeling and flow (source: Philips, 1997).

Table 6.7a. How often do the action steps stand out in software manuals* (n=52)?

	Total	%
Numbered action steps, run-in text	2	3.8
Numbered action steps, indented text	41	78.8
Bulleted list	6	11.5
Other font or signaling of font	2	3.8
Not signaled	1	1.9

* *Two manuals combined two modes of signaling*

Table 6.7b. And in hardware manuals* (n=52)?

	Total	%
Numbered action steps, run-in text	2	3.8
Numbered action steps, indented text	39	75
Bulleted list	5	9.6
Other font or signaling of font	1	1.9
Frame around the actions	1	1.9
Not signaled	4	7.7

* *Three manuals combined two modes of signaling*

To set echo effect:

1. Type **SET-ECHO.**

2. Type <Enter>.

Figure 6.33. Basic action steps (source: Creative Technology, 1994 slightly adapted version).

reflects the characteristics of the objects themselves. When keys and menu objects are capitalized (figure 6.34), the designer follows suit, which is in line with isomorphism. Next, there is an about even preference for capitalizing the complete object (22% vs. 25%) and for presenting it boldface (18% vs. 27%).

2. Click the Capture button.

3. Set up the image you want to capture.

4. Press F7.

Figure 6.34. The first letter of the object in the action step is capital-ized. In this case this coincides with their real world appearance (source: Creative Technology, 1994).

2. Press the **Trim Tool** button.

Figure 6.35. One of the ways of depicting symbols, icons, buttons of keys is to position these in a left- or right-hand column next to the action step. Thus, the name of an object and its picture are con-nected (source: Avid Technology, 1996).

1. Select Install from the system menu.

Figure 6.36. A basic action step with facilitating modifier (source: Pacific Image Communications, 1997).

On facilitating modifiers

One of the elaborations for the basic verb–noun action step involves the presence of a "facilitating modifier." Such modifiers are often locator phrases that direct the user to the place of the object (figure 6.36). The presence of such a facilitating modifier is rather common in the instructions for use, especially in the software manuals. Twenty software manuals and 14 hardware manuals supported users with some form of locator phrase.

The use–order principle of Dixon (1982, 1987) suggests that the preferred order is to present the action step before the modifier. The majority of findings nicely accords with this principle. Thirteen (65%) software manuals and 11 (79%) hardware manuals always used an action–modifier sequence. One software manual was inconsistent with varying modifier–action and action–modifier presentations.

A basic action step with or without facilitating modifier is relatively easy to understand. Complexity increases rapidly when designers include more than a single action and when additional information is

Table 6.8a. What is the preferred form of signaling a key object in the action steps in software manuals (n=49)?

	%
Different type of font	13
Signaling of font	69
Use of symbols	5
Quotation marks	14

Table 6.8b. What is the preferred form of signaling a key object in the action steps in hardware manuals (n=31)?

	%
Different type of font	9
Signaling of font	71
Use of symbols	5
Quotation marks	15

given. Figures 6.37 and 6.38 show some of these more complex forms. In these cases, easier presentation formats and guidance in the form of signals are called for to reduce cognitive load and the risk of error.

On the presentation of keys or screen elements in the action steps

To support the handling of keys or screen elements such as icons and buttons, designers can present these visually or textually. The general design guideline here is isomorphism. Whenever possible, designers should describe or depict the object as it really is. Within a textual format there is a choice between naming the key or screen element (i.e., button), or to add the suffix "key" or "button." The latter form is a preferred choice for novices to prevent them from accidental typing. Of course, the key or button can also be depicted.

Tables 6.9a and 6.9b show the distribution for these three ways of representing keys or buttons. The findings for software manuals clearly show that most designers choose the least preferred option, that of

> 1. From the main Harvard Graphics window, select **File|New presentation** or press **Ctrl-N**. The Add Slide dialog box appears.

Figure 6.37. Complexity increases when more actions are presented in a single action step and when additional information is given. In this case, the additional information contains a facilitating modifier, a shortcut and feedback (source: Harrison, & Yu, 1992).

> Regular disk maintenance is a three-step process described as follows. Note that the order of the steps is important.
>
> As you know, you should make regular backups of your hard disk. Avoid performing the other steps listed here until you have a recent and complete backup set so that any mistake made during disk maintenance can be corrected. The most common error is accidentally deleting an important file, and the Recycle Bin provides some safety from that. Still, the Bin is not permanent and not foolproof, regular backups are your only guarantee.

Figure 6.38. Complexity increases when instructions are more like narratives that must be read. The user is instructed to make a backup and it is assumed that the user knows exactly how to do that (source: Tiley, 1995).

Table 6.9a. How are keys represented in software manuals (n=22)?

	%
Visualized key	9
The name of the key plus 'key' (e.g., Enter key)	14
Just the key name itself (e.g., Enter)	77

Table 6.9b. How are keys represented in hardware manuals (n=24)?

	%
Visualized key or button	17
The name of the key plus 'key' (e.g., Enter key) or The name of the button plus 'button' (e.g., Operate button)	63
Just the key or button name itself (e.g., Enter)	21

naming the key or button without mentioning that this is a key or a button. Maybe this is based on the assumption that this format does not confuse the user. Perhaps the simplicity of the description is seen as an advantage over consistently adding the suffix "key" or "button" (here designers might also consider using a fading technique). For hardware manuals the situation is strikingly different. Although here too the keys or buttons are not depicted very often, the majority of designers add the suffix "key" or "button" after the description.

Conclusion

When we set out to examine the nature of procedures we had not expected the task complexity that we faced. We quickly realized that it would be futile to attempt to give full coverage. From the multitude of factors that impact on the content and format of procedures in software and hardware manuals we have chosen to describe some of what we believe are the most pertinent ones. In doing so, we have let outselves be guided by theories and by what we found. By advancing design guidelines we have attempted to bridge the gap between theory and practice. We see these guidelines as an aid. They are subordinate to a fundamental view on designing technical documentation. Such a view could be the use and user-centered approach that we advocate. Within such an approach the role of procedures is vital and we have tried to contribute our insights as to what these roles might be, how they may be shaped, and how they currently are expressed.

Acknowledgments

The authors want to thank David Farkas and Jean Pratt for their constructive comments on the drafted version of this chapter.

References

ANSI (1998). Z535.1-5. *Accredited Standards Committee on Safety Signs and Colors*. Washington, DC: National Electrical Manufacturers Association.

Alpert, S. R., Singley, M. K., & Carroll, J. M. (1995). Multiple multimodal mentors: Delivering computer-based instruction via specialized anthropomorphic advisors. *Behaviour and Information Technology, 14*, 69-79.

Arnold, W. A. (1988). Learning modules in minimalist documents. *Proceedings of the 35th International Technical Communication Conference (ITCC)*, pp. WE 16-19. Washington, DC: Society for Technical Communication.

Ausubel, D. P. (1968). *Educational psychology: A cognitive view*. New York: Holt, Rinehart & Winston.

Bhavnani, S.K., Reif, F., & John, B.E. (2001). Beyond command knowledge: identifying and teaching strategic knowledge for using complex computer applications. *Proceedings of the CHI'01 Conference on Human Factors in Computing Systems*, pp. 229-236. New York, NY: Association for Computing Machinery.

Bhavnani, S. K., & John, B. E. (1997). From sufficient to efficient usage: An analysis of strategic knowledge. *Proceedings of the CHI'97 Conference on Human Factors in Computing Systems*, pp. 91-98. New York: Association for Computing Machinery.

Bhavnani, S. K., & John, B. E. (2000). The strategic use of complex computer systems, *Human Computer Interaction, 15*, 107-137.

Brockmann, R. J. (1990). *Writing better computer user documentation. From paper to hypertext.* Version 2.0. New York: Wiley.

Brockmann, R. J. (1998). *From millwrights to shipwrights to the twenty-first century.* Creskill, NJ: Hampton Press.

Bruck, B. (1999). *The essential Office2000 book.* Roseville, CA: Prima Publishers.

Card, S. K. , Moran, Th. P. , & Newell, A. (1983). *The psychology of human-computer interaction.* Hillsdale, NJ: Lawrence Erlbaum Associates.

Carroll, J. M. (1990). *The Nurnberg Funnel: Designing minimalist instruction for practical computer skill.* Cambridge, MA: MIT Press.

Carroll, J. M. (Ed.). (1998). *Minimalism beyond the Nurnberg funnel.* Cambridge, MA: MIT Press.

Carroll, J. M., Mack, R. L., Lewis, C. H., Grischkowsky, N. L., & Robertson, S. P. (1985). Exploring exploring a word processor. *Human-Computer Interaction, 1*, 283-307.

Carroll, J. M., & Van der Meij, H. (1998). Ten misconceptions about minimalism. In J. M. Carroll (Ed.), *Minimalism beyond the Nurnberg funnel* (pp. 55-90). Cambridge, MA: MIT Press.

Chandler, P., & Sweller, J. (1991). Cognitive load theory and the format of instruction. *Cognition and Instruction, 8*, 293-332.

Deutsches Institut für Normung (1988). DIN 8418. *Benutzerinformation. Hinweise für die Erstellung* [User guides. Guidelines for their construction]. Berlin, Germany: Beuth Verlag.

Dixon, P. (1982). Plans and written directions for complex tasks. *Journal of Verbal Learning and Verbal Behavior, 21*, 70-84.

Dixon, P. (1987). The structure of mental plans for following directions. *Journal of Experimental Psychology: Learning, Memory and Cognition, 13*(1), 18-26.

Duffy, T. M. , Palmer, J. E., & Mehlenbacher, B. (1992). *Online help design and evaluation.* Norwood, NJ: Ablex.

Farkas, D. K. (1999). The logical and rhetorical construction of procedural discourse. *Technical Communication, 46*(1), 42-54.

Feinberg, S. (1989). *Components of technical writing.* New York: Holt, Rinehart and Winston.

Fleming, M., & Levie, W. H. (1993). *Instructional message design. Principles from the behavioral and cognitive sciences*. Englewood Cliffs, NJ: Educational Technology Publications.

Floreak, M. J. (1989). Designing for the real world: Using research to turn a target audience into real people. *Technical Communication, 36*(4), 373-381.

Frantz, J. P. (1994). Effect of location and procedural explicitness on user processing of and compliance with product warnings. *Human Factors, 36*, 532-546.

Frase, L. T. (1981). Writing, text, and the reader. In C. H. Fredericksen & J. F. Dominic (Eds.), *Writing: The nature, development, and teaching of written communication: Vol. 2: Writing: Process, development and Communication*. Hillsdale, NJ: Lawrence Erlbaum Associates.

Friedmann, K. (1988). The effect of adding symbols to written warning labels on user behavior and recall. *Human Factors, 30*(4), 507-515.

Gellevij, M., Van der Meij, H., De Jong, T., & Pieters, J. M. (1999). The effects of screen captures in manuals. A textual and two visual manuals compared. *IEEE Transactions on Professional Communication, 42*(2), 77-91.

Gellevij, M., Van der Meij, H., De Jong, T. & Pieters, J.M. (2002). Multimodal versus unimodal instructions in a complex learning context. *Journal of Experimental Education, 70*(3), 215-239.

Hartley, J. (1994). *Designing instructional text*. (3rd ed.). London: Kogan Page.

ISO (1983). ISO 37. *Instructions for use of products of consumer interest*. Geneva: International Organization for Standardization.

ISO (1984). ISO 3864. *Safety colours and Saftey Signs*. Geneva: International Organization for Standardization.

ISO (1988). ISO 3461-1. *General principles for the Creation of Graphical Symbols*. Geneva: International Organization for Standardization.

Jennings, K. (1990). *The devouring fungus: Tales of the computer age*. New York: Norton.

Jones, M., & Myers, D. (1988). *Hands-on HyperCard*. New York: Wiley.

Krull, R. (1994). Documentation for a physical world. *Journal of Technical Writing and Communication, 24*(2), 181-195.

Lathrop, L. (2001). Index estimator. Available online at: www.indexingskills.com/tips.html.

Laughery, K. R., & Hammond, A. (1999). Overview. In M. S. Wogalter, D. M. DeJoy, & K. R. Laughery (Eds.), *Warnings and risk communication* (pp. 3-13). Philadelphia, PA: Taylor & Francis.

Lazonder, A. W. (2001a). Efficacy of minimalist instruction to develop self-regulatory skills to search the Web. Paper presented at the ninth European Conference for Research on Learning and Instruction (EARLI), Fribourg, Switserland.

Lazonder, A. W. (2001b). Minimalist instruction for learning to search the World Wide Web. *Education and Information Technology, 6*(3), 161-176.

Leonard, S. D., Otani, H., & Wogalter, M. S. (1999). Comprehension and memory. In M. S. Wogalter, D. M. DeJoy, & K. R. Laughery (Eds.), *Warnings and risk communication* (pp. 149-188). Philadelphia, PA: Taylor & Francis.

Leutner, D. (2000). Double-fading support—A training approach to complex software systems. *Journal of Computer Assisted Learning, 16,* 347-357.

Lorch, R. F., & Chen, A. H. (1986). Effects of number signals on reading and recall. Journal of Educational Psychology, *8*(4), 263-270.

Mack, R. L., Lewis, C. H., & Carroll, J. M. (1990). Learning to use word processors: Problems and prospects. In J. Preece & L. Keller (Eds.), *Human-computer interaction: Selected readings* (pp. 185-204). Hemel Hempstead: Prentice Hall.

Maes, A., Maas, A., Van der Meulen, I., & Verbunt, F. (1998). *Wanneer waarschuwen waarschuwingen?* [When do warnings warn?] Taalbeheersing, *20*(2). 126-140.

Martin-Michiellot, M., & Mendelsohn, P. (2000). Cognitive load while learning with a graphical computer interface. Journal of *Computer Assisted Learning, 16,* 284-293.

Mayer, R. E. (1987). *Educational psychology. A cognitive approach.* Boston: Little, Brown.

Mayer, R. E. (1999). Designing instruction for constructivist learning. In C.M.Reigeluth, *Instructional-Design theories and models. Vol. II: A new paradigm of instructional theory.* Mahwah, NJ: Lawrence Erlbaum Associates.

Mayer, R. E., & Gallini, J. K. (1990). When is an illustration worth ten thousand words? *Journal of Educational Psychology, 82,* 715-726.

Mirel, B. (1998). Minimalism for complex tasks. In J. M. Carroll (Ed.), *The Nurnberg funnel: Designing minimalist instruction for practical computer skill* (pp. 179-218) Cambridge, MA: MIT Press.

Newell, A., & Simon, H. A. (1972). *Human problem solving.* Englewood Cliffs, NJ: Prentice-Hall.

Paivio, A. (1990). *Mental representations: A dual coding approach.* New York: Oxford University Press.

Price, J., & Korman, H. (1993). *How to communicate technical information. A handbook of software and hardware documentation.* Redwood City, CA: Bejamin/Cummings.

Smith, P. L., & Ragan, T. J. (1999). *Instructional design* (2nd ed.). New York: Wiley.

Strawbridge, J. A. (1986). The influence of position, highlighting and imbedding on warning effectiveness. *Proceedings of the Human Factors Society 30th Annual Meeting* (pp. 712-715). Santa Monica, CA: Human Factors Society.

Sweller, J. (1994). Cognitive load theory, learning difficulty, and instructional design. *Learning and Instruction, 4,* 295-312.

Sweller, J., & Chandler, P. (1994). Why some material is difficult to learn. *Cognition and Instruction, 12*, 185-233.

Sweller, J., Van Merrienboer, J., & Paas, F. G. W. C. (1998). Cognitive architecture and instructional design. *Educational Psychology Review, 10*(3), 251-296.

Thomas, R. C., & Foster, M. R. K. (2001, Jan/Feb). *A pilot study of teaching the strategic use of common computer applications.* Paper presented at the Australian User Interface Conference, Bond University, Australia.

University of Chicago Press. (1982). *The Chicago manual of style* (13th ed.). Chicago, IL: University of Chicago Press.

Van der Meij, H. (1992). A critical assessment of the minimalist approach to documentation. *Conference Proceedings of the 10th Annual International Conference on Systems Documentation,* SIGDOC92, Ottawa, Canada (pp. 7-17). New York: ACM.

Van der Meij, H. (1994). Catching the user in the act. In M. F. Steehouder, C. Jansen, P. van der Poort, & R. Verheijen (Eds.), *Quality of technical documentation* (pp. 201-210). Amsterdam/Atlanta: Rodopi.

Van der Meij, H. (1996a). A closer look at visual manuals. *Journal of Technical Writing & Communication, 26*(4), 371-383.

Van der Meij, H. (1996b). Does the manual help? An examination of the problem-solving support offered by manuals. *IEEE Transactions on Professional Communication, 39*(3), 146-156.

Van der Meij, H. (2000). The role and design of screen images in software documentation. *Journal of Computer Assisted Learning, 16*, 294-306.

Van der Meij, H., Blijleven, P. J., & Jansen, L. M. (2001). *Codeboek voor het analyseren van procedures in software – en hardware handleidingen* [Codebook for the analysis of procedures in software and hardware manuals]. Internal Report. Enschede, the Netherlands: Twente University.

Van der Meij, H., & Carroll, J. M. (1998). Principles and heuristics for designing minimalist instruction. In J. M. Carroll (Ed.), *Minimalism beyond the Nurnberg funnel* (pp. 19 -53). Cambridge, MA: MIT Press.

Van der Meij, H., & Gellevij, M. (1998). Screen captures in software documentation. *Technical Communication, 45*(4), 529-543.

Venema, A. (1989). *Produktinformatie ter preventie van ongevallen in de privésfeer. Gevaars- en veilighiedsinformatie op verbruiksprodukten* [Product information for the prevention of accidents in the home and during leisure activities: Hazard and safety information on non-durable products] . Research Report 69. Leiden, the Netherlands: Institute for Consumer Research, SWOKA.

Weiss, E. H. (1991). *How to write usable user documentation* (2nd ed.). Phoenix, AZ: Oryx Press.

Wiedenbeck, S., Zila, P. L., & McConnell, D. S. (1995). End-user training: An empirical study comparing on-line practice methods. *CHI'95*, 74-81.

Wiedenbeck, S., Zavala, J. A., & Nawyn, J. (2000). An activity-based analysis of hands-on practice methods. *Journal of Computer Assisted Learning, 16*(4), 358-365.

Wogalter, M. S. (1994). Factors influencing the effectiveness of warnings. *Proceedings of Public Graphics* (5.1-5.21). Lunteren, the Netherlands.

Wogalter, M. S., DeJoy, D. M. & Laughery, K. R. (1999a). Organizing theoretical framework: A consolidated communication-human information processing (C-HIP) model. In M. S. Wogalter, D. M. DeJoy, & K. R. Laughery (Eds.), *Warnings and risk communication* (pp. 15-23). Philadelphia, PA: Taylor & Francis.

Wogalter, M. S., DeJoy, D. M., & Laughery, K. R. (Eds.). (1999b). *Warnings and risk communication*. Philadelphia, PA: Taylor & Francis.

Wogalter, M. S., & Leonard, S. D. (1999). Attention capture and maintenance. In M. S. Wogalter, D. M. DeJoy, & K. R. Laughery (Eds.), *Warnings and risk communication* (pp. 123-148). Philadelphia, PA: Taylor & Francis.

Figure References

Avid Technology (1996). Avid Videoshop 3.0. Getting started manual. Tewksbury, MA: Avid Technology Inc.

Carroll, J. M., & Van der Meij, H. (1998). Ten misconceptions about minimalism. In J. M. Carroll (Ed.), Minimalism beyond the Nurnberg funnel (pp. 55-90). Cambridge, MA: MIT Press.

Corel (1997). Corel Magic Gallery. Canada: Corel Corporation.

Cowart, R. (1998). Het complete boek Windows '98 internet editie. Soest, Nederland: Sybex..

Creative Technology (1994). Sound Blaster 16, User Reference Manual. Singapore, Creative Technology.

Evers, B. (1994). Dubbelboek Lotus 1-2-3 Versie 5 (NL). Soest, Nederland: Sybex.

Gardner, D.C., & Beatty, G.J.(1994). Visuele leermethode voor WP6 voor Windows. Utrecht, Nederland: Bruna MultiMedia.

Gravis (1994). Gravis Utilities User's Guide. San Mateo, CA: Advanced Gravis Technology.

Hanson, L. (1993). Everything you always wanted to know about the Mac. Haydon Books.

Harrison, D., & Yu, J.W. (1992). Harvard Graphics for Windows: Self Teaching Guide. New York, NY: Wiley.

Heid, J. (1992). MacWorld guide to Microsoft Word 5. San Mateo, CA: IDG Books Worldwide.

Hewlett-Packard (1996). Creatief afdrukken. Palo Alto, CA: Hewlett-Packard Company.

Lazonder, A. W. (2000). Leren werken met Internet Explorer. Internal report. Enschede, Nederland: Universiteit Twente.

Le Blanc, D. A., & Le Blanc, R. (1995). Using Eudora: The user-friendly-reference. Indianapolis, IN: Que Corporation.

Microsoft (1991). *Introductie Handboek Microsoft Works*. Seattle, WA: Microsoft Corporation.

Pacific Image Communications (1997). Supervoice 2.2 for Microsoft Windows, User Manual. England, Pacific Image Communications.

Philips (1997). Philips Brilliance 107 17" Digital Autoscan Color Monitor. Taiwan: Philips.

Samsung (1996). Samsung CD-ROM DRIVE User's Guide. Korea: Samsung.

Sony (1994). CD-ROM DRIVE Unit-User's Guide. Japan: Sony Corporation.

Stuur, A. (2000). Windows 98 voor senioren. Utrecht: Bruna MultiMedia.

Sweller, J., & Chandler, P. (1994). Why some material is difficult to learn. *Cognition and instruction, 12*, 185-233.

Tae il Media (1996). CD-ROM Drive CDD-7120. Tae il Media Corporation.

Tiley, E. (1995). Windows '95 unleashed. Indianapolis, IN: Sams Publishing.

Unknown (1996). 1454 Color Monitor User's Guide.

Valentine, C.S. (1993). Inside Access 1.1, Special edition. Carmel, CA: New Riders Publishing.

WordPerfect Corporation (1993). WordPerfectPresentaties. Versie 2.0 Naslag. Orem, UT: WordPerfect Corporation.

Zardetto Aker, S. (1991). The Macintosh Companion, The Basic and Beyond. New York: Addison Wesley.

7

Visual Design Methods in Interactive Applications

Jean Vanderdonckt
Université catholique de Louvain

Visual design in general is the arrangement of information items (e.g., text, images, diagrams, pictures, tables) in such a way that the resulting product is visually attractive, perceptive, and easily understandable. Visual design issues are raised in many domains of human activity such as user interface design, documentation development, presentation design, and graphic layout. This chapter describes techniques coming from traditional visual design and discusses them in the context of user interface design.

When the designer sketches the components of a user interface, she or he first selects appropriate interaction and interactive objects according to the user's task (Vanderdonckt & Bodart, 1993). The second activity is to determine the basic layout of these selected objects ranging from the most important to the least important: the main application window, the title and menu bars, the functional areas of the application window (e.g., a status bar, a toolbar), all child windows, dialog boxes and panels with their contents.

This layout (figure 7.1) consists of interaction objects and interactive objects. Interaction objects (IO), also called widgets or controls, encompass static objects (e.g., labels, separators, group boxes) and dynamic objects (e.g., edit boxes, radio boxes, option boxes). Interactive objects cover every kind of object that a multimedia user interface could display: static icons, drawings, pictures, images, sketches, video sequences, graphics and so forth. Each of these objects allows some special interaction with the user. For instance, an image of the human body may include hot spots for defining different sensitive regions of the body in order to be selected, displayed, explained, or zoomed. Some images can be extracted from a video sequence in order to be analyzed. Interaction and interactive objects are henceforth referred to as IO.

Determining a basic layout consists of calculating and drawing any geographical composition of functional areas of the user interface into a comprehensive format depending on the user's task. In particular, solving the layout problem for a dialog box consists of drawing the set of related IO, assembling them into a rectangular area, and surrounding them by borders. The layout then looks like a set of rectangles when drawn around each IO (figure 7.1a). A *layout grid* consists of a set of parallel horizontal and vertical lines that divide the layout into units that have visual and conceptuel integrity (de Baar, Foley, & Mullet, 1992; Feiner, 1991; Marcus, 1992; Taylor, 1960). The intersections of these lines delimit these units into rectangles that constrain the IO position (figure 7.1b). Equally spaced lines typically establish external margins in the layout and consistent space between the different IO. Layout grids are very practical for form fill-in user interfaces and for text displays since their IO reflect the layout of the source doument or the page of a book (Hurlburt, 1978; MüllerBrockman, 1981). Such a layout grid can be applied for both the background and foreground of screens, as in HyperCard (Apple, 1992).

The problem comes from the fact that in modern user interfaces, these grids are no longer as useful because the layout no longer consists of vertical and horizontal lines. Instead of these kinds of lines, the layout may be based on other lines (e.g., oblique lines, discontinuous lines), convex shapes (e.g., lozenge), planes (e.g., a plane with a vanishing point), and volumes (e.g., cylinder). Therefore, we extend the definition of a layout grid to include a *layout frame*. A layout frame consists of dots, lines, shapes and volumes that constraint the localization of IO. Deciding such a complex layout frame is not an easy task. To help the designer do this job, several visual techniques are now described.

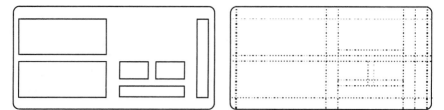

Figure 7.1. (a) A layout of a dialog box; (b) The underlying layout grid.

Definition of visual techniques

A *visual technique* relies on a commonly accepted visual principles to suggest the arrangement of the layout frame components. The visual techniques listed in this section are sorted by similarity and not by rank of importance because we are convinced that all visual principles cannot be applied with the same representativeness. Some principles are very easy to apply, some other are more difficult to deal with, and some other become very hard to translate. Moreover, applying this or that principle mostly depends of the involved IO and the visual aims that the designer has in mind. Visual techniques can be grouped into five categories:

- *Physical techniques*: balance, symmetry, regularity, alignment, proportion, and horizontality.
- *Composition techniques*: simplicity, economy, understatement, neutrality, singularity, positivity, and transparency.
- *Association and dissociation techniques*: unity, repartition, grouping, and sparing.
- *Ordering techniques*: consistency, predictability, sequentiality, and continuity.
- *Photographic techniques*: sharpness, roundness, stability, levelling, activeness, subtlety, representation, realism, and flatness.

These techniques are discussed further in the next sections.

Physical techniques

Balance is a highly recommended technique evoqued by many authors (Dondis, 1973; Dumas, 1988; Galitz, 1989; Horton, 1990; Kim & Foley, 1993). Balance is a search for equilibrium along a vertical or horizontal axis in the layout. If a weight is attached to every IO, balance requires that the sum of IO weights on each haxis remains similar. *Instability* is the opposite of balance where IO are not distributed equally on the axis. "They seem ready to topple over" (Galitz, 1989). Balanced layouts are not only easy to understand, but, also, easy to design by a game of counterpoise. If an IO is placed to the left of the vertical axis, instability is provoked and immediately countered by adding an IO of the same weight to the right of the vertical axis.

Balance does not have to take the form of symmetry. Balance can be realized through symmetry and asymmetry. Symmetry provides a balance to the layout by centering titles, headings on both sides of the axis, by placing two columns of equal length, one on the left, one on the right. The weights of IO can be adjusted asymmetrically (Dondis, 1973), although it is technically more complicated to reach a balance with dynamic asymmetry than with static symmetry. Reaching asymmetric balance is a matter of weight, size, and position.

Symmetry (Dondis, 1973; Galitz, 1989; Kim & Foley, 1993) consists of duplicating the visual image of IO along a horizontal and/or vertical axis (e.g., left on the right, top to bottom, or vice versa). Achieving symmetry automatically preserves balance, but the balance can be performed without symmetry. Symmetry is very simple to verify and logical to imagine, but can lead to static layouts without originality (Dondis, 1973). The opposite of symmetry is *asymmetry* where at least one IO does not possess a replication on the other side of the axis.

Regularity (Dondis, 1973; Galitz, 1989) is a visual technique establishing uniformity of IO placed according to some principle, method, convention that does not change from one layout to another. For instance, a layout where IO are uniformly spaced in columns and rows is described as regular. *Irregularity* occurs when no such principles, method or convention exists, when no logical order of IO is apparent. Irregularity emphasizes unexpected, unusual, and unconforming layout grids (Dondis, 1973).

Alignment (Kobara, 1991; Mullet & Sano, 1995; Streveler & Wasserman, 1984) is probably the most accessible and practical visual technique. Alignment occurs when the number of vertical alignment points in a row and the number of horizontal alignment points in a column is reduced or minimized. *Misalignment*—the opposite of the alignment—occurs when the number of alignment points is greater than one. Misalignment is accentuated when all IO containing task's data are placed just after their identification labels. Fonts with descenders and ascenders may affect alignment of similar IO if badly used.

Proportion (Dondis, 1973; Galitz, 1989; Kim & Foley, 1993; Marcus, 1992; Mullet & Sano, 1995; Tufte, 1983) describes an aesthetically appealing ratio between the dimensions of IO (often composite IO). Because dimensions exist in the real world, we can feel them, we can see them, we can compare them. The ratio is calculated by dividing the height of an IO by its length. Several proportions have been either proved aesthetic (e.g., the Golden Ratio) or widely and conventionnally preferred (e.g., $1{:}\div 2$, 1:2, 1:1.29, 1:1.5, 1:4/3, 1:1.6 as recommended by Marcus (1992) and Tufte (1983)). *Disproportion*—the opposite of proportion—is implied at the time no special ratio is used or a large difference appears between the two dimensions.

Horizontality (Dondis, 1973) is a corollary of the previous technique: layouts with greater length than height (i.e., a horizontal ratio) are predominant. *Verticality*—the opposite of horizontality—occurs if layouts have greater height than length (i.e., a vertical ratio). Horizontality is generally preferred over verticality in user interface design.

Composition techniques

Simplicity (Dondis, 1973; Galitz, 1989; Mullet & Sano, 1995; Tullis, 1981; Tullis, 1983) is directness and singleness of layout, free from secondary complications or sophistications. Simplicity often improves the ease of understanding in a layout. Simplicity is guaranteed by placing IO according to a logical and natural arrangement (e.g., by frequency, by physical property) driven by the task's semantics. *Complexity*—the opposite of simplicity—increases visual intricacy with too many objects and hinders any organization of the layout grid. Tiled IO are considered a simple layout; varying overlapping IO are considered a complex layout.

Economy (Bowman, 1968; Dondis, 1973; Galitz, 1989) is the frugal and judicious use of IO in the layout to present information as simply as possible. Economy can be pursued when necessary and sufficient IO are placed in the layout, and nothing else: no IO that are extraneous to the user's task. The aim of economy is the fundamental visual layout, emphasizing the conservatism and understatement of the poor and the pure (Dondis, 1973). Economy is intended to define the boundaries of necessity within which it can work sucessfully (Bowman, 1968). *Intricacy*—the opposite of economy—occurs when unfrequent, unwanted IO encumber the layout, visually or not. This situation particularly occurs when highly detailed or digitized images with a lot of decoration are placed rather than simple IO that are reduced to the essentials and whose important features are the only salient ones.

Understatement (Dondis, 1973) and its opposite, exaggeration, are equivalent to the couple economy/intricacy, but in the domain of intellectual, mental representation rather than physical, spatial representation. Understatement supposes that the viewer of the layout is able to deduce a maximum of information from a minimum of IO to be presented. The verbal counterparts of understatement is euphemism and ellipsis (i.e., the art of saying many things with few words). *Exaggeration* shows in the layout a minimum of information with maximized IO. The verbal counterpart of exaggeration is hyperbole. Exaggeration is achieved through extravangancy, amplified expressions that are enlarged far beyond possible.

Neutrality (Dondis, 1973; Horton, 1990) A neutral atmosphere in the layout is obtained by placing all IO at the same level, with the same presentation attributes (or, at least, with little variations) as much as possible and, preferably, with no highlighting method (e.g., no blinking, no underlining, no bolding, no boxing). *Accent*—the opposite of neutrality—is equivalent to the rendering of any highlighting method on a particular IO against a sameness of background. Most graphical highlighting methods are useful: reverse video, color, brightness, boldness, boxes, borders, different sizes, overprinting, magnifying (Horton, 1990).

Singularity (Dondis, 1973) is the focus of a layout on one separate and solitary IO, unsupported by any other IO or composiion of IO. Specific emphasis is conveyed on a simple IO, despite the presence of other IO. *Juxtaposition*—the opposite of singularity—expresses an interaction between IO placed side by side or to be compared with an activated relationship or to be related by any visual technique.

Negativity (Dondis, 1973; Horton, 1990) displays IO in dark colors on a light background. Black IO (text, separators, labels, files) and colored IO (bitmaps, images) are generally displayed on a white or grey layout. *Positivity*—the opposite of negativity—displays IO in a bright color on a dark background. If negativity has been experimentally tested to reduce errors and reading time, to increase subjective satisfaction and legibility, positivity may still be used to convey special atmospheres, often with light IO (e.g., grey pictures).

Transparency (Dondis, 1973; Staples, 1993) means a visual layout where IO, superseded by other IO, can still stay visible behind or through them. Transparency is typically mandatory when displaying text on a colorful picture. A light transparent surface (e.g., grey or blue) is added between the text and the picture to improve the legibility of the text and to allow, in spite of all the visibility, the picture behind the text. *Opacity*—the opposite of transparency—means the complete blocking out, concealing of IO that become visually occulted. Having partially occulted IO force the user to guess what IO (part or complete) are hidden. Opacity can also be used to give the impression of a relative distance or depth (see depth below) legislated by overlapping.

Association and dissociation techniques

Unity (Bowman, 1968; Dondis, 1973; Galitz, 1989) is the placement of individual IO into one totality (e.g., a window) that is visually all of a piece. With unity, all IO seem to belong to each together and to be bound so that they can be seen as a whole and taken as one sealed unsectile thing: Seeing one element is seeing the whole. Unity can be revealed with similar sizes, colors, surrounding blank spaces, logical organization exhibiting interrelation of IO in terms of the whole. With *fragmentation*, all IO seem to be isolated, to retain their own character themselves.

Repartition (Dumas, 1988; Horton, 1990) occurs when IO are shared mong the four quadrants of the layout as systematically as possible. *Quadrant preference*—the opposite of repartition —occurs when IO are preferably placed in one or many specific quadrants. Of course, we have taken into account the fact that human eyes favor the lefthand and lower area of any layout (this phenomenon is called *Preference for lower left*). But there are numerous examples of poorly distributed user interfaces. Most of these examples show displays in which IO are pushed over to the lefthand portion of the layout.

Is it because Western users read from left to right or because programmers find very easy this way of placing IO? In all cases, the repartition should be compatible with the task structure, rather than with the file or database structure.

Grouping (Dondis, 1973; Galitz, 1989; Horton, 1990; Tullis, 1981; Zahn, 1971) is a visual technique that creates a circumstance of give and take of relative interaction. Grouping is mainly based on the *law of attraction*: Two grouped IO fight for attention in their interaction by establishing individual statements depending on the distance between the IO. The closer the IO are, the stronger the attraction is. Grouping is also affected by the *law of similarity*: When dissimilar IO are grouped, the human eye increases the relation between them. When similar and dissimilar IO are grouped, hidden connections are identified quickly. Grouping is one of the best techniques for structuring a layout, namely by providing an aesthetic appearance, by helping remembering, and by accelerating a layout search. One of the most important usability guidelines is the fact that objects that are semantically linked should be grouped in some way (Vanderdonckt; 1994). Contrarily, objects that do not share any semantic relationship should be split. *Splitting*—the opposite of grouping—means that no such structure is visible: IO are displayed without the ability to visually perceive an attraction or a repulsion between IO.

Sparing (Dondis, 1973; Galitz, 1989; Horton, 1990) looks to avoid cluttered or overcrowded layouts: It suggests keeping the visual loading of a layout within reasonable boundaries. *Density*—the opposite of sparing—does not consider whether IO are stacked and packed too tightly in the layout. Generally, many layouts contain so much IO that easy scanning is impossible. The trend is to fill each layout space with as much IO as possible (e.g., text, fields, pushbuttons, images). The *visual loading*, sometimes called density, is, by definition, the proportion of busy positions on the layout. For alphanumeric displays, it can be expressed as the ratio of displayed characters by the total amount of characters in the layout (Tullis, 1983). In graphical user interfaces, the density is calculated by dividing the number of lighted pixels by the total number of available pixels. Streveler and Wasserman (1984) also measured the field density, which is the total number of fields (static or dynamic) in the layout, and the box density, which is the total number of visual groups whether surrounded or not. Tullis (1983) recommended that layout density should not exceed 25%. Horton (1990) suggested that the density of a welldesigned paper page be about 40%.

Ordering techniques

Consistency (Dondis, 1973) is a visual technique for expressing visual compatibility with the subject, for developing a layout whose IO are dominated by one sound, uniform, constant thematic. Consistency takes place not only in the dimensions or the ordering of IO, but also in their (little) differences. *Variation*—the opposite of consistency—is the strategy for identifying changes, elaborations as variations in musical themes. Variations do not necessarily take the form of unconsistency, where the same IO are laid out at different places from one layout to another. Moreover, variation can be assumed by a series, a continuum of IO whose contents, shapes, colors, and themes vary significantly.

Predictability (Dondis, 1973; Galitz, 1989) is a visual technique where IO are placed according to some order or plan that is highly conventional and recognizable. Knowing the information structure of the task, viewing one layout or remembering it should enable the user to predict how another layout will be arranged. Predictability also suggests the user is able to foretell in advance what the entire layout will be just by seeing a minimum part, or some significant part, of it. Predictability is enhanced through layout consistency. *Spontaneity*—the opposite of predictability—does not suggest such a highly conventional plan. The user will therefore be unable to infer successive layouts from already viewed layouts or to generalize the entire layout from its parts.

Sequentiality (Dondis, 1973; Galitz, 1989) is a layout plan that is arranged in a logical, rythmic, expected order. Many orders can be followed to sequentially place IO: numerical order, alphabetical order, chronological order, physical order, type order, sequential order, functional order, logical order, frequence order, importance order, consensus order, designation order, and so on. *Randomness*—the opposite of sequentiality—promotes the absence of a particular ordering plan, that is, a layout where the IO flow cannot be detected due to the lack of plan, or a disorganized, accidental, random one.

Continuity (Dondis, 1973) keeps the visual connections between the IO. These connections are fundamental for preserving a unitized visual statement. Continuity can be achieved by uninterrupted steps from one IO to another. This is often the case in video sequences, series of snapshots, morphing pictures, and so forth. Continuity means the cohesion of the parts to the whole layout. *Episodicity*—the opposite of continuity—interrupts the visual connections that may exist between the IO. Episodicity means the reinforcement of the parts of a layout without adding new meaning or message with all IO taken as a whole.

Photographic techniques

Sharpness (Dondis, 1973) is a visual technique that is closely related to clarity of both physical state and expression. Having sharpness in a layout can be interpreted as having:

- Clearly distinctive IO (e.g., information to be input should be distinctive of information to be displayed by the system).
- IO with precise outlines, hard edges, distinct margins.

The effect of sharpness is a distinct atmosphere. *Diffusion*—the opposite of sharpness—opts for less precision of character, more fuzzy, more feeling and warmth.

Roundness (Dumas, 1988; Horton, 1990) is the preference for curved IO. *Angularity*—the opposite of roundness—is the preference for IO with angular, rugged outlines.

Stability (Dondis, 1973) is the expression of preference for IO that have clear foundation. *Stress*—the opposite of stability—occurs when IO are not placed on their firm base or stability: A circle is a good example. Placing a rectangle or a triangle on one of its corners causes stress. When an IO is intrinsically irregular, the analysis and establishment of balance is more involved and intricate.

Leveling is a visual technique for automatically establishing balance through artifacts. IO are laid out so that balance axis will stand out. Through perception, balance can be emphasized (or rubbed out, repectively) when we recognize easily (with difficulty, respectively) the abstract visual condition of balance. This is often the case when IO are equally distributed in two colums with two alignment points per column. *Sharpening*—the opposite of leveling—on the other hand, destroys any automatic balance by placing IO on unexpected, unbalanced locations.

Activeness (Dondis, 1973) reflects motion through explicit representation or implicit suggestion. The goal of activeness is to design an active and energetic layout with lively postures (e.g., arrows, stopped image in a video sequence, an action snapshot). *Passiveness*—the opposite of activeness—withdraws any IO that could bring a dynamic effect. Passiveness relies on the technique of static representation, which presents an atmosphere of quiescence, resting by equilibrating IO.

Subtlety (Dondis, 1973) is a visual technique used in order to make a fine distinction, shunning any obviousness and energy of purpose. Subtlety is often synonymous with ingeniosity since it requires delicate, highly refined IO. *Boldness*—the opposite of subtlety—looks for every obvious IO in its context. Boldness is often synonymous with optimum visibility of all IO in the layout.

Representation (Dondis, 1973) subsumes subtletly because its intended purpose is to use IO that concretely represent the real world in details. *Abstraction*—the opposite of representation—uses IO that abstract the real world in many ways. Icons, for example, can be representative if they simply translate a physical object or they can be abstract if they mimic some action or just represent metaphors or major characteristics of the physical objects. For instance, concrete icons are believed to be better than abstract icons.

Realism (Dondis, 1973) is the natural technique of camera. Many tricks and conventions are able to replicate the same visual cues that our eyes convey to our brain when receiving an external image. Realism tries to follow the camera in the same way by reproducing exactly what we see (e.g., "what you see is what you lay out"?). Perspective (Staples, 1993) is one possible technique for providing realism in 3D contexts. *Distortion*—the opposite of realism—tampers with realism, seeking control of effect through the deformation of the real IO in shape, form, or color. This technique covers zooming in and out, magnifying lens, fisheye views (Furnas, 1986; Leung & Apperley, 1993), pictures cut in moved rows, torn pictures. Distortion is an eyecatching way to produce an intense response.

Flatness (Dondis, 1973) does not use any technique for providing perspective, so erasing the natural feeling of dimension and space. *Depth*—the opposite of flatness—tries to render perspective by replicating the environment through effects of light, shade, gradient of color, overlapping (Seligmann & Feiner, 1991).

Examples of visual techniques

Physical and ordering techniques are typically the visual techniques that the designer may apply for the layout of traditional applications, since these techniques are effective for alphanumeric displays and for graphical user interfaces. This is not the case with real multimedia applications. Real multimedia applications are not those database applications where one or two fields containing images or video have just been added (in this case, rectangular layout grids are still useful). Figure 7.2 depicts a simple multimedia application with its related grid. Real multimedia applications are those applications where the screen consists of a real mix of images, pictures, video, text, graphics controls, icons, and so on, with a high degree of interaction, not just leaving the user to gaze passively.

Figure 7.3 illustrates the three first sets of visual techniques (i.e., physical, composition, association and dissociation techniques). Symmetry is not preserved, but balance is partially established with the two columns whose length are equal, but whose heights differ significantly. The plan used for regularity promotes the localization of the picture on the upper left part of the layout, because this information is the center of all interactions and informations displayed elsewhere. Alignment is extremely well achieved because of the two columns reducing the number of alignment points. Even if these columns highlight more verticality than horizontality, they are sized with appropriate proportions: The area containing the picture has a good aspect ratio (Feiner, 1988), the textual zones are limited in height and separated.

The background color of this layout is mainly black, whereas textual information is displayed with white and green as foreground color. The x-ray picture is in grayscale. Therefore, positivity with a very high degree (this may obscure the reading) is reached rather than negativity, but the accent is placed on the picture because it is lighter. Repartition is guaranteed by filling the two columns of the layout, with pushbuttons on the bottom. Grouping is also seen via the progressive disclosure of objects: The title and the picture are first displayed, textual information follow group by group, with different definitions highlighted in each group, commands end the reading path of the layout in each column.

The order of objects in each column is highly predictable and sequential, following a natural and logical order of reading and understanding. Locating the text objects first should force the reader to first look at the picture on the right before understanding the contents of the information.

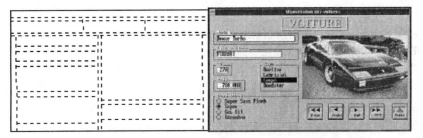

Figure 7.2. A simple multimedia layout and its related grid.

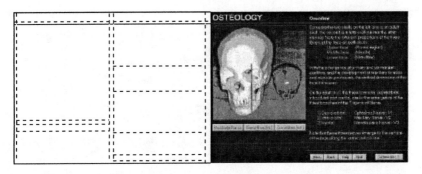

Figure 7.3. A multimedia layout in osteology and its related grid.

Photographic visual techniques are relevant to the picture: sharpness, angularity; stability, leveling, passiveness, boldness, representation, realism and flatness give a 6/9 score in favor of harmony and a 3/9 score in favor of contrast.

Despite these qualities, this screen still relies on a layout grid only consisting of vertical and horizontal lines. Although they are not numerous and very simple, they do not suggest a very sophisticated interaction between the system and the user. All physical visual techniques (e.g., symmetry, balance) are respected in the case of figure 7.4. This is a good example of achieving vertical symmetry (in the centre of the second column) and balance independently of the contents of the pictures. Other characteristics of figure 7.4, although very similar to figure 7.3, are very economic, neutral, and, moreover, continuous. The rectangular pictures are equally shaped, following the western convention of reading from the left to the right, and from the top to the bottom. Each picture represents a particular state in the life of the bones. A visual continuity is then established between the images, which is further reinforced by full juxtaposition.

Obvious outline gives sharpness in the space of the work environment of a user (figure 7.5). Stability is apparent in this layout since the main window and the wires of the space are horizontal and vertical. A slight sharpening is provoked by moving the window on the left rather than on the center, but this provides more the feeling of space. Activeness is raised by the impression of movement toward the vanishing point. Subtlety, distortion, and depth are all produced by the illusion of space and perspective. The wired representation of perspective is more abstract than a real space representation, but this representation becomes more obvious to the eyes.

Figure 7.6 highlights that a sequence of screen can take advantage of reusability of the underlying grid. Indeed, the same top area with the aircraft and the takeoff/landing strip is repeated on each layout and

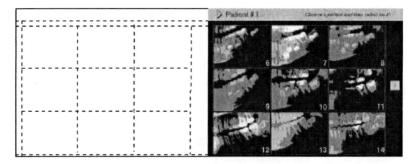

Figure 7.4. A continuous multimedia layout and its related grid.

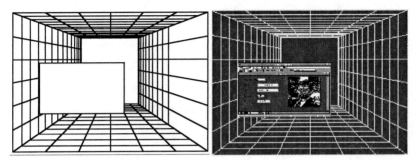

Figure 7.5. A layout of a work environment and its related frame.

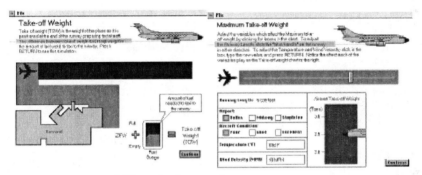

Figure 7.6. Reusability of a layout frame.

different bottom areas are presented according to the nature of information. These two screens prove that consistency may be partially applied, though the rest of the layouts prefer variation. This apparent lack of consistency does not necessarily reduce other ordering techniques from the time the user remarked the different portions.

Layout guidelines based on visual techniques

The experience gained by applying these techniques suggests some guidelines, as already done in HyperCard (Apple, 1992):

- Apply physical visual techniques for traditional layouts, especially when the layout is text-dominant rather than graphic-dominant; balance is the most important visual technique to achieve;

- Apply composition visual techniques with contrast so that visibility is not endangered;
- Apply association and dissociation visual techniques with contrast carefully: Fragmentation and quadrant preference may hold in certain cases, but splitting and density are to be avoided in every case;
- Apply ordering visual techniques with contrast where the user's task is intrinsically unstructured or asynchroneous, and with harmony where the user's task is structured or sequential;
- Apply exaggeration, accent, singularity to draw a visual impact on the most important objects of the frame if no ordering visual techniques can be applied;
- Apply photographic visual techniques for pure multimedia elements only;
- Prefer abstraction rather than representation in each frame where the real world should not necessarily be depicted as is: Abstract images or raster pictures are more easily understood than complex digitized images;
- Combine visual techniques with contrast only when appropriate; the combination of techniques increases the visual impact of the frame, but uncareful combination may destroy the intended purpose.

Conclusion

By summing up techniques used in each category, it is possible to rank screens by comparison. Visual techniques per se do not mean that a screen oriented toward harmony is better than a screen oriented toward contrast or vice versa. Rather, it helps designers to more precisely estimate the impact of such a screen, depending on the target audience. If a "traditional" audience is expected, perhaps with efficiency goals in mind, it is likely that visual techniques in the domain of harmony will be preferred. Instead, if a "nontraditional" audience is targeted, visual techniques in the domain of contrast may catch more attention than the traditional ones. Furthermore, they may also increase the visual cognitive load. This decision is a tradeoff between usability and desired visual impact.

Although our body of knowledge about multimedia objects is still insufficient today, several points can be highlighted:

- Multimedia objects can be qualified with certain attributes (Faraday & Sutcliffe, 1993; Sutcliffe & Faraday, 1994).
- Reasonable proportions for displaying images and video sequences are already known (Feiner, 1991).

- Visualization techniques for special objects (3D objects, maps) are well studied (Furnas, 1986).
- The layout frame should be governed by these attributes and informations.

In addition, it is conceivable that the aforementioned visual techniques could be included as part of a tool to help designers generate better multimedia layouts.

Additional study of such visual techniques in the realm of multimedia objects can only improve designs. All multimedia objects exhibit these techniques; the only real question is whether they are the ones the designer intended.

Note

For additional material related to this chapter, please see www.isys.ucl.ac.be/bchi/members/jva/pub/visual.htm.

References

Bodart, F., & Vanderdonckt, J. (1993). Expressing guidelines into an ergonomical styleguide for highly interactive applications. In S. Ashlund, K. Mullet, A. Henderson, E. Hollnagel, & T. White (Eds.), *Adjunct Proceedings of ACM Conference on Human Factors in Computing Systems INTERCHI'93* (pp. 35-36). New York: ACM Press.

Bowman, W. J. (1968). *Graphic communication.* New York: John Wiley & Sons.

de Baar, D., Foley, J. D., & Mullet, K.E. (1992). Coupling application design and user interface design. In P. Bauersfeld, J. Bennett, T. G. Lynch (Eds.), *Proceedings of the Conference on Human Factors in Computing Systems CHI'92* (pp. 259-266). New York: ACM Press.

Dondis, D. A. (1973). *A primer of visual literacy.* Cambridge, MA: The MIT Press.

Dumas, J. S. (1988). *Designing user interfaces for software.* Englewood Cliffs, NJ: Prentice Hall.

Faraday, P., & Sutcliffe, A. (1993). Toward a walkthrough method for multimedia design. In G. Salvendy (Ed.): *Proceedings of HCI International'93* (pp. 452457). Amsterdam: Elsevier Science.

Feiner, S. (1988). A gridbased approach to automating display layout. In *Proceedings of Canadian Annual Conference on Graphics Interface GI'88* (Edmonton, Alberta, Canada, June 610, 1988), Canadian HumanComputer Communications Society, 1988 (pp. 192-197).

Feiner, S. (1991). An architecture for knowledgebased graphical interfaces. In J. W. Sullivan & S. W. Tyler (Eds.), *Intelligent user interfaces* (pp. 259-279). New York: ACM Press.

Feiner, S., & McKeown, K. (1990). Coordinating text and graphics in explanation generation. In *Proceedings of AAIA90* (pp. 442-449).

Foley, J., van Dam, A., Feiner, S., & Hughes, J. (1990). *Fundamentals of interactive computer graphics*. Reading: Addison Wesley.

Furnas, G. W. (1986). Generalized fisheye views. In M. Mantei & P. Orbeton (Eds.), *Proceedings of ACM Conference on Human Factors in Computing Systems CHI'86* (pp. 16-23). New York: ACM Press.

Galitz, W. O. (1989). *Handbook of screen format design*. Wellesley: Q.E.D. Information Sciences.

Horton, W. K. (1990). *Designing & writing online documentation helpfiles to hypertext*. New York: Wiley.

Hurlburt, A. (1978). *Layout: The design of the printed page*. New York: WatsonGuptill Publishing.

HyperCard® stack design guidelines (1990), Apple Computer Inc. Reading, MA: AddisonWesley.

Kim, W. C., & Foley, J. D. (1993). Providing highlevel control and expert assistance in the user interface presentation design. In *Proceedings of ACM Conference on Human Factors in Computing Systems INTERCHI'93* (pp. 430-437). New York: ACM Press.

Kobara, S. (1991). *Visual Design with OSF/Motif*. Reading: AddisonWesley.

Leung, Y. K., & Apperley, M.D. (1993). E3; Towards the metrication of graphical presentation techniques for large data sets. In *Proceedings of International Conference on HumanComputer Interaction EWHCI'93* (pp. 9-26), Vol. II.

Marcus, A. (1992). *Graphic design for electronic documents and user interfaces*. New York: ACM Press.

MüllerBrockman, J. (1981). *Grid systems in graphic design*. Niederteufen: Arthur Niggli Verlag.

Mullet, K., & Sano, D. (1995). *Designing Visual Interfaces: Communication Oriented Techniques*. Upper Saddle River, NJ: Prentice Hall.

Seligmann, D. D., & Feiner, S. (1991). Automated generation of intentbased 3d illustrations. In T. W. Sederberg (Ed.), *Proceedings of SIGGRAPH'91* (pp. 123-132).

Staples, L. (1993). Representation in virtual space: Visual convention in the graphical user interface. *Proceedings of the Conference on Human Factors in Computing Systems INTERCHI'93* (pp. 348-354). New York: ACM Press.

Streveler, D. J., & Wasserman, A. I. (1984). Quantitative measures of the spatial properties of screen designs. In B. Shackel (Ed.), *Proceedings of the First IFIP TC13 Conference on Human-Computer Interaction INTER-ACT'84* (pp. 1125-1133). Amsterdam: Elsevier Science.

Sutcliffe, A., & Faraday, P. (1994). Designing presentation in multimedia interfaces. In *Proceedings of ACM Conference. on Human Factors in Computing Systems CHI'94* (pp. 92-98). New York: ACM Press.

Taylor, I. (1960). Perception and design. In J. Ball & F. C. Pyres (Eds.), *Research principles and practices in visual communication* (pp. 51-70). Bloomington: Association for Educational Communication and Technology.

Tufte, E. R. (1983). *The visual display of quantitative information.* Cheshire, MA: Graphics Press.

Tullis, T. S. (1981). An evaluation of alphanumeric, graphic and color information displays. *Human Factors, 23*, 541-550.

Tullis, T. S. (1983). The formatting of alphanumeric displays: A review and analysis. *Human Factors, 25*, 657-682.

Vanderdonckt, J. (1994). *Guide ergonomique des interfaces hommemachine [Usability guide of user interfaces].* Namur: Presses Universitaires.

Vanderdonckt, J., & Bodart, F., (1993). Encapsulating knowledge for intelligent automatic interaction objects selection. In S. Ashlund, K. Mullet, A. Henderson, E. Hollnagel, & T. White (Eds.), *Proceedings of ACM Conference on Human Factors in Computing Systems INTERCHI'93* (pp. 424429). New York: ACM Press.

Zahn, C. T. (1971). Graphtheoretical methods for detecting and describing gestalt clusters. *IEEE Transactions on Computers, X20*, 68-86.

8

Contextual Inquiry as a Method of Information Design

Karl L. Smart
Central Michigan University

As we enter what has been called the information age, the value and importance of creating, managing, and using information effectively has increased. In many ways, information has become the ultimate commodity—a profitable product line for businesses (Passini, 2000, p. 84) As the importance of information has grown, the field of information design has emerged. Unlike some related more established disciplines, information design still lacks a comprehensive foundational theory, with limited research and case studies to demonstrate its effectiveness or to provide empirical guidelines for practitioners (Jacobson, 2000, p. 2). This chapter responds, in part, to this paucity in research by reviewing a user-centered approach to information design—contextual design—through a case study that demonstrates how a large computer software company applied the contextual design process to the information and documentation components of their software with the intent of improving the experience of users.

The chapter begins with a discussion of the importance of customer experience and the need for customer input in product and information development, followed by a discussion of the contextual design methodology. The application of the methodology is shown, demonstrating how the software company sought to enhance users' experience with their software through better supporting learning and problem-solving strategies. The suggested design innovations that resulted from the study are discussed, exploring the impact of innovation on the proposed information components of the software. The chapter concludes by discussing some of the constraints affecting the implementation of design ideas along with implications for further research and practice.

The importance of customer experience

Some claim that information design is a recent iteration of an "age-old profession of communications assistance" (Horn, 2000, p. 2). In its current context, however, information design specifically involves preparing information for efficient and effective use by individuals, including appropriate interactions with computers and other equipment as well as accurate, comprehendible, usable documents. Building upon the business tradition of designing products and services to meet customer needs and expectations, information designers have come to understand the need to present "the right information to the right people at the right time, in the most effective and efficient form" (Horn 2000, pp. 15-16).

Information design—whether signage or a computer interface or documentation—often involves assisting individuals in completing a task, be it reaching a destination, solving a problem with a computer application, or assisting in the use of some product. An implicit foundation of information design involves providing appropriate information to the individual trying to use that information for accomplishing a goal. In business terms, such support or assistance results in customer satisfaction.

Although assisting or satisfying customers is significant in and of itself, it has larger implications. Seeking customer satisfaction has become an important business strategy many companies use to gain competitive advantage and maintain economic viability (Parasuraman, 1997, Rust & Zahorik, 1993). As Norman (1989) observed,

> Modern industry must distinguish itself through its consider-
> ation of the needs of its customers. . . . As companies design
> more for usability and understanding, they will discover a
> competitive edge, for these principles save customers time and
> money while increasing morale. (pp. vi, vii)

Increasingly, companies seek customer satisfaction through creating experiences for customers, knowing that meaningful experiences create added value (Norton & Hanson, 2000; Pine & Gilmore, 1999).

Organizations, however, have come to realize they cannot rely on designers, developers, or specialists to intuitively know how to design products and services to create meaningful customer experiences or to meet customer needs. Designers bring their own biases, rationalizations, and views to development efforts, which often interferes with assessing what customers truly need or want. Research shows that the more customer contact a project has the more likely it is to be successful (Keil & Carmel 1995).

With the increasing complexity of technology, designers need a corresponding increase in the awareness of customer needs and how humans interact with systems. As Hackos and Redish (1998) observe, "Good design happens only when designers understand people as well as technology. ... Designs that don't meet users' needs will often fail in the workplace or in the market, resulting in high costs in productivity, frustration, and errors that impact users and their organizations..." (pp. 1, 25). Although an increasing amount of research has been published on user-centered product development, especially in the computer industry (den Buurman, 1997; Leonard & Rayport, 1997; Carroll, 1995; Schuler & Namioka, 1993), less research demonstrates the application of a user focus to information design (Raven & Flanders, 1996; Beabes & Flanders, 1995).

This chapter provides an example of how a user-centered design process was adapted by a computer software company to enhance the information experience of software users. As an interpretative case study, the chapter shows how a development team within the company worked with users to explore how users learned new software and how they behaved when encountering problems within software applications. Specifically, the team explored methods of user assistance and documentation that assisted users in these tasks, with the goal of finding improved ways of assisting users and improving their experience with software applications.

The case study demonstrates contextual design methods refined by Hugh Beyer and Karen Holtzblatt (1998) to gather and analyze information on users wants, needs, and works habits when using computers. I begin by explaining the methodology of contextual design, outlining the development team organization and research focus. I show where and how a development team gathered information from users and how the team conducted subsequent interpretative sessions to organize the data and to discover insights about users and user tasks. After discussing various diagrams and work models developed from the data, I review some of the methods the team used to generate innovative design ideas and to prototype and test preliminary ideas among users. I conclude by discussing constraints that affected the implementation of the design ideas, with implications for practice and implications for additional research.

Contextual design methodology

Beyer and Holtzblatt (1998) defined contextual design as "a field data-gathering technique that studies a few carefully selected individuals in depth to arrive at a fuller understanding of the work practice across all customers" (p. 37). As noted, significant research has focused on customer-centered design methods for computer applications. However, dealing with the information components or documenting the software has traditionally been viewed as a secondary support issue, with less time and effort devoted to creating documentation and support truly based on customer need. With a growing awareness of the need to define documentation more broadly than merely online help and a printed manual, the software company of this study organized a development team to explore the communication and information components of software.

Team organization and research focus

Because the emphasis of the study involved information and documentation, the company organized the leadership of the team from the documentation department. A documentation group manager was selected to lead the team, with others from the documentation department assigned full time to work on the team: a technical editor (familiar with print), a technical writer (known as the online help "guru"), and a technical writer who had been an editor. Knowing that the documentation department could not independently effect the changes that would likely result from the study, the team recruited others from outside documentation to join them in creating a more cross-functional group needed for greater validity, acceptance, and implementation. Subsequently, the team added a lead software developer, a usability specialist, a UI designer, and several other software developers that rotated through the course of the project.

With abundant complaints as to the uselessness of documentation, particularly in the popular press (Irvin, 1996; Grech, 1992; Rettig, 1991), the team (designated as the Contextual Inquiry or CI team) wanted not only to determine how documentation is used but how users learn and interact with applications and how best to support that interaction. As a result, the team set out initially not to design a new style or type of documentation but to better understand users—their work practices and needs—and to determine how the applications developed, not just the documentation, could best assist users and improve their experience with software. This focus led the CI team to specifically look at how users learned new software, how they got "unstuck" within computer applications, and how documentation (define broadly as any information in the program, from menu structures and long prompts to printed manuals and online help) interfaced with their learning and problem-solving strategies.

The CI team realized that information designers and technical communicators frequently work from many assumptions about how and why individuals use or do not use documentation but that few of these assumptions had been verified. The team's work attempted to find out the work structure of individuals using software and how documentation supports and assists users in completing their work. The major focus—How do people learn a software program and how do users get unstuck?—centers on two primary experiences when documentation supports a user. This focus was used when team members conducted the actual site interviews with users.

Gathering user data

Following Beyer and Holtzblatt's (1998) suggestion of interviewing 15 to 20 customers from at least four to six work sites (p. 156), the CI team identified potential organizations and individuals for the study. To make certain the interviews captured a heterogeneous group of users, the team selected a representative number of male and female users from different types of organizations in different regions of the United States. The team conducted a total of 17 interviews at 10 different sites, interviewing 7 men and 10 women. Four of the interviews were conducted with individuals working at government agencies, eight in businesses, and five at educational institutions—the sites being located in five regions of the United States: Seattle, Washington; San Francisco, California; Salt Lake City, Utah; Chicago, Illinois; and Durham, North Carolina.

One or two members of the team went to every interview, usually one to ask questions and work with the user and one to take detailed notes. In addition, each interview was tape-recorded. The person taking the notes would later listen to the tape recording and validate the notes taken. Part of the data gathering included collecting artifacts from the site—such things as copies of pages referred to in print documentation, a printed copy of online help screens, and copies of any internal documentation or user-created tipsheets. Additionally, the team gathered detailed notes about the work culture and environment.

The interview team watched users using a variety of software (not just the software developed by the company): word processors, presentation graphics, spreadsheets, email, and custom-built in-house applications. The intent of the interviews centered on observing users actually working. To get users performing tasks in the context of their work environment, interviewers posed questions that forced users to do not just recollect their work: When was the last time you were stuck in an application? What did you do? How did you get unstuck? Can you show us the process by recreating what you did? Such questions helped get users actually working and as a result revealed work practices frequently hidden or forgotten by the user. The average interview lasted 3 to 4 hours.

Interpreting the data

After collecting data from the interviews, the interview teams returned from the sites to meet with the entire CI team to interpret the data. Interpretative sessions were held as soon as possible after the interview teams returned from a site visit. The night before an interpretive session, the interview team would listen to the tape recording of the interview and review their notes.

As the CI team gathered, one person sat at a computer and captured information in note card format as the interviewers discussed the interview. Each note card created identified the user (U1, U2, etc.) and sequence of the information in the interview (1, 2, 3, etc.). Each card contained a single bit of information from the interview or a possible design idea or insight, as shown in figure 8.1. The note cards were displayed during the meeting so the team could verify and correct the information. The cards became a permanent record of the interview and were later used in creating an affinity diagram that organized the data.

In addition to the note cards, members of the team drew up to five different work models capturing information about the interview. Typically, the following five work models were created for each user: a sequence model, a flow model, a cultural model, an artifact model, and a physical model. Table 8.1 briefly describes each model and its purpose.

Examples of consolidated models created from the individual work models are shown in subsequent figures. The models captured details about the users' work environment and various aspects of customers' work practice. Along with the note cards, the work models became the basis from which the team built an affinity diagram. Formalizing the notes and models ensured that later discussions were based on actual data, and that design ideas and discussions could be grounded from actual data. The CI team followed this process of note taking and model creating in interpretative session for each of the 17 users interviewed, producing a total of 1,407 note cards.

Creating an affinity diagram

After capturing information from the interviews on note cards and in work models (which occurred over a series of months), the CI team followed an inductive process of bringing all the data together. The first step involved creating an affinity diagram. Affinity diagrams are based

U8 #562	U8 # 563
I'm stuck. I don't know what to do. I can't find the previous window. Tries to find the window by starting over.	He relies on the long prompts to navigate. He doesn't know how to use common Windows navigation
U8 #564	U8 #565
Gets back to dialog box. "These buttons don't work." Sees the dialog box in online Help and thinks that it is a real dialog box.	Design idea: Don't duplicate the interface in online Help. Users confuse it as the real thing. Consider using a break-away portion of the box or have Help take you to the real box.
U8 #566	U8 #567
Chooses "About" from the Help menu expecting to find information about his problem. "About" displays copyright information, etc.	Question: Are people confused when applications use their own viewers to provide Help? Should applications use the standard environment Help?

Figure 8.1. User Information Captured on Note Cards in Interpretative Sessions. (Note cards can include actual quotes, description of actions, interviewer or team questions, and potential design ideas.)

on quality principles and processes developed in Japan (Brassard, 1989; Kawakita, 1982). The intent of the diagram is to organize data across all the customers. The process helps reveal the scope of the data and identifies any holes or weaknesses in the data. Organizing the data within an affinity diagram makes key issues stand out, emphasizing key knowledge about the customer in an easy-to-share format (Beyer & Holtzblatt, 1998). The process also helps the team arrive at a consensus as to what the data mean.

The affinity diagram was built bottom-up, finding common themes and structures from the individual note cards. For example, a member of the team would select a note card that identified a particular issue or idea, and then other team members would look for additional note cards that captured similar information (or had an "affinity" or relationship to the original card). After the notes were grouped, the team created a name or succinct phrase to represent the group of cards. After the team named all the groups, they then looked for relationships and structures among the groups. This resulted in a hierarchical structure that organized the data into manageable chunks.

Table 8.1. Description of Different Work Models Created in
Interpretative Sessions

Model Type	Description and Purpose
Sequence Model	Captures the process of a work task or action, including a trigger that causes the sequence, steps involved in the process (with corresponding order), and any breakdowns in the steps.
Flow Model	Shows the individual user in relationship with others and the flow of information between and among individuals. Captures the flow of information, artifacts used, communication topics and actions, places where work is done, and breakdowns.
Cultural Model	Captures the culture of the user's work environment, including expectations, desires, policies, values, influences, and attitudes about work.
Artifact Model	Provides an actual representation of artifacts in the user's environment that are involved in the user's work, with explanatory notes keyed to how the artifact is used.
Physical Model	Depicts the actual physical environment of the user (physical structure, arrangement of furniture and objects, layout of work and artifacts).

Figure 8.2 shows a portion of the affinity diagram developed from the study.

The first level of the affinity diagram is a major group label identifying one of the focuses of the study. The second level lists one of the major subheadings that summarize a set of groups. The third level (e.g., I go to a book when I get stuck) shows group labels that reflect the common affinity of groups of individual note cards. The individual note cards appear below each group title. Additional initial user strategies for getting unstuck include the following:

- Try to figure it out myself, even though I know something could help me do it faster.
- Try some standard trouble-shooting strategies (even when I don't understand the problem).

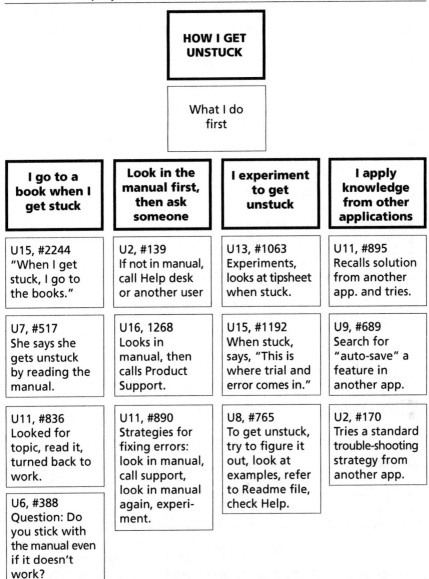

HOW I GET UNSTUCK			
What I do first			
I go to a book when I get stuck	**Look in the manual first, then ask someone**	**I experiment to get unstuck**	**I apply knowledge from other applications**
U15, #2244 "When I get stuck, I go to the books."	U2, #139 If not in manual, call Help desk or another user	U13, #1063 Experiments, looks at tipsheet when stuck.	U11, #895 Recalls solution from another app. and tries.
U7, #517 She says she gets unstuck by reading the manual.	U16, 1268 Looks in manual, then calls Product Support.	U15, #1192 When stuck, says, "This is where trial and error comes in."	U9, #689 Search for "auto-save" a feature in another app.
U11, #836 Looked for topic, read it, turned back to work.	U11, #890 Strategies for fixing errors: look in manual, call support, look in manual again, experiment.	U8, #765 To get unstuck, try to figure it out, look at examples, refer to Readme file, check Help.	U2, #170 Tries a standard trouble-shooting strategy from another app.
U6, #388 Question: Do you stick with the manual even if it doesn't work?			

Figure 8.2. Abbreviated Portion of the CI Team's Affinity Diagram

- Check for simple things first.
- Remember information from training.
- Continue with faulty hypothesis.
- Talk to myself as I think through the problem.
- Go to other people or places for information.
- Try things I'm told, even when I don't think they'll help.

The affinity diagram helped the team develop a hierarchical structure for the data, providing organization to the knowledge and insight gleaned by the team. It provided a single structure to organize the customer data and revealed some of the common issues and themes as well as the scope of problems and issues users encountered. Along with the work models, the affinity diagram showed key elements of work practices and became a crucial foundation for identifying design requirements. The completed affinity diagram covered the entire wall of a large conference room and served as a constant reference for supporting decisions. When members of the CI team would make a claim about a user or design idea, they frequently referred to the diagram to validate their claim. The affinity diagram allowed the CI team to begin to envision design and documentation solutions from particular problems users encountered. The diagram became the support or rationale for innovative design ideas, allowing the team to develop ideas in response to the data.

Consolidating work models

After creating the affinity diagram, the team set about to consolidate the work models of each of the users. The purpose was to create concise visual representations of and statements about the customer population, showing common structures without losing variations across customers (Beyers & Holtzblatt, 1998). Through the models, the CI team created "maps" of the customer population, their work practices, and environment. What follows is a brief description of the consolidated models and some of the key issues identified with each.

Sequence model

Although the affinity diagram reflected specific thoughts, values, and behaviors, it did not represent the specific process of various actions in the way a sequence model did. As the CI team interviewed users, the teams captured detailed notes on specific tasks that users performed. Consolidating individual sequence models demonstrated specific task structures and showed strategies common across the user-population.

For example, one consolidated model captured what users do with information they find to assist them in learning or getting unstuck, detailing the process by showing steps users take, identifying breakdowns that occur, and indicating jumps to other sequences. This sequence model showed various strategies users employ when they have to locate information in documentation and identified potential breakdowns. In looking at redesign, the team analyzed each of the breakdowns and looked at ways the system or the documentation could assist users who encountered problems. This consolidated sequence model along with other models are discussed further in a subsequent section on design issues.

The sequence model shows various strategies users employ when they have locate information in documentation and identifies potential breakdowns. In looking at redesign, the team analyzed each of the breakdowns and looked at ways the system or the documentation could assist users who encountered problems. This consolidated sequence model along with other models are discussed further in a subsequent section on design issues.

Flow model

As noted, flow models show an individual user's relationship with others within the organization, with the flow of information between individuals. Consolidating the flow models of users in the study revealed basic patterns of communication and work practices. It showed how organizations defined various user jobs and roles. As the CI team analyzed and consolidated flow models, they discovered ways that applications and software could assist users in performing their jobs.

For example, the CI team wanted to analyze the flow of information between a knowledge worker or user and others within the organization. Specifically, where did a user get information for solving a problem and were there any breakdowns in that information. One flow model captured how information flowed between a users and others within as well as outside the organization

Cultural model

Cultural models reflect the culture of a user and the relationships among individuals within the user's environment. The CI team wanted to show the informal and formal relationships of users to individuals they turned to in getting help or assistance. In conjunction with a flow model, the CI team identified three categories of information assistance: informal internal, formal internal, and formal external. The cultural aspects of this model indicates the relationship of the user to others within and without the organization. One of the most significant implications from cultural model is that information was more costly and less reliable the further it was from the user.

Artifact model

Many of the artifacts the interview teams collected were pages from manuals (both company-produced user's guides that shipped with products and third-party manuals purchased individually) and printed copies of help screens. Additionally, the team saw a frequent use of user-created tipsheets, documentation created informally to provide help or assistance to specific problems. The intent of the artifact models was to see similarities and differences in the documentation users found helpful. The team created consolidated artifact models of each of these artifact types: user's guides, third-party manuals, Windows Help screens, and user-generated tipsheets.

With the user's guides, for example, the consolidated artifact model showed commonalties among several user's guides along with several third-party manuals. In addition to commonalties, individual differences were noted. The artifact model of the software companies' user guides and third-party manuals showed that organizationally and informationally (the actual content) third-party manuals were almost identical to manuals that shipped with the product. In analyzing the differences between the two types of documentation, however, the CI team found that rhetorically the third-party manuals used a chattier, more friendly tone, which in part may be responsible for their greater perceived usefulness.

Besides manuals, tipsheets proved to be a common artifact collected from users. Tipsheets suggested that users could not find solutions to particular work problems in documentation (or that they did not have ready access to documentation), and as a result they create their own documentation for future reference and to share with others within the organization solutions to problems encountered. Considering the importance and prevalence of tipsheets to users, the CI team looked for ways applications could support the generation and distribution of tipsheets.

Physical model
The consolidated physical model reflected the general physical environment of the users interviewed. Although the team observed differences among the worksites visited, they found striking similarities. A central focus of the office was a computer at the work desk, with a phone, books and reference material, and a work area close to the computer.

The physical model confirmed data from the flow model, which shows that users often seek assistance from colleagues or others in close proximity to their own work space. It also showed the availability of help through easily reached manuals or telephone support, whether to a product's support line or a company help desk.

The physical model, along with the other consolidated models and the affinity diagram, captured critical data about users that the CI team used for design purposes. The following are some of the key insights the team gleaned from the interpreted data:

- Users create their own documentation.
- The interface is part of the documentation.
- Indexes are crucial entry points to documentation.

- Quicktips are important in helping users complete tasks and get unstuck.
- Users need things explained or referenced in their terms (they need to be able to point to the thing they want).
- Users do not care if their methods are inefficient.
- Users get angry when a new version "trashes" their knowledge base by "improving" the software.
- Error-message recovery is very important.
- Users have little time for formal learning or training.
- Tutorials are seldom used in the real world.
- Some users—such as answer givers or macro programmers—want or need a printed manual.

Each of these insights was substantiated from actual data in the consolidated models and affinity diagram. The following section discusses how the team used this data to generate design ideas and what implications these insights had for information design.

Using data to inform design decisions

A challenge with any user-centered design approach is finding ways to transform collected data into actual design ideas. Recent research has focused on ways to bridge the gap between information and actual design (Isensee, Pierce, & Righi, 2001; Wood, 1998). With the contextual design methodology, the CI team followed a process of visioning, building a user environment and storyboarding, and prototyping actual design ideas. Once they validated ideas through prototyping, the team developed design specifications and worked within technological and organizational constraints to implement their ideas.

Visioning

Realizing that they were not just designing a new type of manual or help system, the CI team tried to look "out of the box" to come up with creative solutions to impact a user's experience. The gathered data reflected a diversity of users with different learning styles and needs who used different strategies when they encountered problems. For instance, from the models and affinity diagram, the team identified the following profiles of learning styles among users:

- Experimenters (users who like exploring on their own).
- Model Builders (users who want to see the big picture and build models of how things work).
- Do-to-Learn Users (those who learn by doing).

- Learn-to-Do Users (those who want to learn a concept first, then to try it out).
- Watch-and-Ask Users (who want to see a task completed by someone else).
- Read-and-Follow Users (those who like explicit instructions and will follow directions step by step).
- Accidental or Indifferent Users (users who have little motivation or desire to learn, who learn serendipitously if at all).

Although it seems self-evident that different users exhibit different learning styles and strategies, the differences have significant implications for creating users' experiences. Systems (and the documentation and information within systems) should facilitate various learning strategies and support different problem-solving approaches, depending upon the need or desire of an individual learner. The CI team tried to envision distinct methods of support for the different learning profiles identified.

In addition to identifying learning styles, the CI team discovered several strategies users utilized when encountering problems:

- Try the same procedure or steps again (and again, even if the same problem resulted).
- Go back to "square one—ground zero" (which usually involved rebooting and starting the whole system fresh).
- Ask a buddy, pod-mate, or anyone within earshot.
- Consult the documentation set to find information.
- Call the Help desk.

With such diversity in learning styles and problem-solving strategies, the team used a "visioning" process to invent possible responses to different types of users and problem-solving strategies. The vision did not involve creating a new tool or document; rather, it focused on what the user needed or needed done, and then looked for appropriate technology to build a system that met the identified needs.

With the wide diversity of users and their accompanying needs, the CI team wanted a system that was aware of users and their differences and that could somehow customize assistance and help according to varying user needs. The team developed the concept, originally called "the Blue Fairy," that would somehow track what users were doing and provide assistance when needed. Realizing that this involved two separate functions, the team refined the idea into what became known as the *Watcher* and the *Communicator*.

The function of the Watcher was to analyze the setup of the user's computer and monitor the user's actions. By monitoring the user, the

Watcher could identify possible problems. For instance, the data from the affinity diagram showed that if users tried the same sequence unsuccessfully three times, they did not understand the problem they had encountered and needed help. The Watcher then would automatically clue in if a user tried to perform an action and repeated the action three times without success. Information gathered by the Watcher would then be transferred to the Communicator.

The Communicator would become the part of an application that directly interacted with users. It would help users understand and operate the application, warn the user of problems, step the user through tasks as needed, and help the user recover from errors. The Communicator could interact with users through voice, typing text, or pointing to pictures or parts of the interface.

From the data, the team realized that one of the primary strategies users turned to for getting information was a fellow worker. The intent of the Communicator was to create an electronic "neighbor" that could provide needed information. For instance, if the Watcher noted that the user had unsuccessfully tried to complete an action three times (an action dubbed "the magic sequence"), the Communicator would intervene and ask, "Are you lost?" The Communicator, based on information from the Watcher, could then suggest possible interventions. The reliability of the information from the Communicator would be high because of its awareness of the user based on the information gathered by the Watcher.

With a Communicator and Watcher in place, the CI team next explored methods of assistance that best suited individual needs and preferences. Table 8.2 highlights some of the major ideas the team developed in providing assistance to users.

In generating this list of ideas, the CI team did not focus on the immediate technical or financial feasibility of the ideas; rather, they realized they needed a vision of what an application could do from which they could implement ideas over a series of product releases. The vision of user assistance then became a basis for building an actual user environment.

Building a user environment design and storyboarding

The involvement of developers became critical in the phase of translating the vision ideas into the reality of a functioning application since many of the concepts required the development of fundamental program features. Many of the vision ideas were only just becoming a possibility with the technological developments and spread of the Internet. The CI team realized that the work had to be long term and ongoing.

Table 8.2. Types of user assistance developed from CI data.

Assistance Type	Description and Purpose
Demos	Demonstrates tasks as users watch. Let's users see task concepts.
Maps	Provides a "big picture" of the system, showing relationships between features and tasks. Gives another way for users to access features and navigate the interface.
More Info	Gives users a way to expand interface items (text and icons) to get more information (such as long prompts expanding to paragraphs of information, with steps or examples as needed).
Cue Cards	Provides users with exact steps to follow online, helping users know exactly what to do. Displays steps according to who the user is, what the user wants, and where the user is in the program.
Tipsheet Manager	Lets the user record steps of an action or task and creates a tipsheet the user can run (as a cue card), save, print, or send to someone else.
Printed "Cookbook"	Gives the user information without having to use the computer. Lets users scan in printed form to follow along online or customize to fit individual needs.
Third-Party Information	Allows the Communicator to access information and support produced by third-party companies.
Communication Center	Provides an area that allows two or more workers to work and talk together asynchronously to solve problems.

For example, the concept of the Watcher and Communicator required applications to be aware of users' actions. Effective use of a Watcher could only work, however, if there was greater system awareness via the operating system. At this point in the development of products, Windows—the primary operating system considered—lacked the necessary awareness of user events. Additionally, the Internet was just coming into play as a new communication and interaction medium. Ideas like the Communication Center and easy access to third-party information needed a majority of users connected so that the system could provide online resources. Moreover, technological constraints still limited some vision ideas, due to processor speeds, hardware, connectivity, and download speeds. Intelligent searching was still limited, and gigabyte storage devices were not common. Nonetheless, the team developed a user environment containing many of these ideas, knowing the implementation would have to occur over several releases.

Figure 8.3 shows an actual initial prototype of the types of user assistance envisioned within the system. The prototype showed how the various types of assistance relate to one another.

With an idea of the users' environment, the CI team set about redesigning several of the consolidated sequences in a storyboard format, integrating new aspects from the new user environment. For example, in a storyboard of a redesigned "Learn to Do" sequence, the user trigger was "I have time to learn." The intent was that users want to orient themselves with the application, deliberately wanting to explore and learn about the application. In the new user environment, a user could go to the Communication Center to access and engage in an online discussion group. Or, the system could generate a list of contact individuals who could provide individual assistance to the user.

If users preferred more immediate interaction, they could go to options within the application itself. If users wanted to get the "big picture" ("Watch-and-Ask" Learners) they could access a Demo, a free standing quick-tour video. Although the data suggested users seldom used tutorials, many users liked watching someone perform an action before trying it themselves. Another way to get an overview would be to explore the application's structure through the Maps feature (useful for a "Model Builder" Learner). Maps are graphical representations that depict the way features and tasks fit together, with possible maps including system maps, task maps, feature maps, comparison maps, menu maps, or button maps.

For example, new users wanting to learn how to create a new document in a word processor could select a task map entitled, "Make a Document." Clicking on options within the map would show users how to use application features to complete the task of creating a document. The map would include a visual index that displayed a finished document, with specific aspects of the document labeled to assist users in using those features—such as borders, bulleted lists, headers and footers, or graphics.

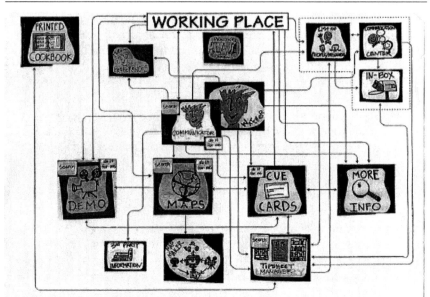

Figure 8.3. Initial Prototype of User Environment (capturing design ideas for user assistance).

The purpose of the storyboarding was to take the vision and integrate it with the consolidated sequences, resulting in a redesign of work practices. The intent was to bring actual work practices in line with the vision. The storyboards visually looked like the consolidated sequence models, with the addition of features from the redesigned user environment integrated into the sequences. The storyboarding provided the first opportunity for the CI team to test their envisioned redesign with actual task sequences. The storyboarding provided an actual structure to the vision and provided the starting point for prototyping the ideas. The prototyping formalized the context for testing aspects of the envisioned user environment.

Prototyping documentation

After the CI team had formalized a new user environment and conceptualized how the environment would work with task sequences and storyboards, they validated the design ideas with actual users through prototypes, an effective and economical way to gather customer feedback on design ideas (Hackos & Redish, 1998; Beyer & Holtzblatt, 1998; Nielsen, 1993). The paper prototypes allowed users to test the new design ideas in scenarios of actual work situations, allowing the team to see if design concepts actually worked.

For example, the CI team created a prototype of a Task Map during their testing. The Task Map displayed tasks users could select, depicted in thumbnail images. If users selected the formal report thumbnail on the left, a larger image of a page from a report appeared on the right, with triangle markers highlighting aspects of the page users may want to create, such as titles and headings, tables, rules, and footnotes. Task maps provides a type of visual index to application features and tasks.

Prototyping design ideas allowed the team to identify additional problems users encountered with preliminary design concepts. The CI team performed numerous iterations of prototypes on each of the major design ideas in the user environment.

Writing specifications

Once the team had adequately tested and refined the design of the user environment, they selected those features that they could work toward integrating in the next product release. With the assistance of several developers, the CI team formalized the user environment and developed corresponding detailed design specifications. Due to financial, technological, and time constraints, the team realized that several of their proposed ideas could not be implemented until future releases, but the formalized user environment with its accompanying design specs became the blueprint from which future development efforts ensued.

For example, the CI team created a prototype of a Task Map during their testing. The Task Map displayed tasks users could select, depicted in thumbnail images (such tasks as a formal report or a memo). If users selected the formal report thumbnail on the left, a larger image of a page from a report appeared on the right, with triangle markers highlighting aspects of the page users may want to create, such as titles and headings, tables, graphic lines, and footnotes. Task maps provide a type of visual index to application features and tasks. Figure 8.4 shows the formalized user environment with its proposed user assistance (next page).

As shown in Figure 8.4, the team assigned each feature a function number (e.g., F 360, F363, etc.) that corresponded with design specifications. The design specs provided more detailed information about the feature, with additional function numbers assigned to each aspect of the feature for development referencing. The design specifications provided a concrete plan for developing software, interface, and documentation requirements for the next release, with the formalized new user environments helping development to help plan for future releases. For example, given technological and time constraints, the concept of the Watcher could only minimally be implement. However, the vision gave developers an idea of what the team planned in subsequent releases.

Dealing with constraints and resistance

As Norman (1989) noted, design ultimately must "take into account real issues in cost, manufacturability, and aesthetics" (p. 3). Ultimately, what the CI team envisioned was more than user assistance: it included a broader concept of the entire user experience where assistance is seamlessly and unobtrusively integrated into the application. As in any organization, due to certain constraints, the entire vision could not be implemented immediately. The two most significant factors affecting the implementation of the vision were technological constraints and corporate constraints and resistance.

An important part of the user experience that the CI team envisioned required an intelligent system that the team described as both pushing and pulling information. The team found that to truly respond to users, a system needed to pull or gather information from users. But in addition, when the system found a user stuck (repeating an action three or more times while unsuccessfully completing a task—the "magic sequence"), the system need to "push" information or assistance to the user.

Several technological constraints prohibited this push–pull of information. For instance, the team still had hardware constraints with processor speed, connectivity, download speeds, and memory (gigabyte storage devices were not common). Additionally, the operating system was not enough aware of user events—a system limitation that applications could not work around. Also, the use of intelligent agents was limited. Moreover, the Internet was just becoming a significant medium in businesses and the work place. For some components of user assistance to work—such as the Communication Center—the application required a majority of users to be connected so that online resources and data could be seamlessly retrieved from the Web. These types of constraints kept many aspects of the vision from being implemented immediately.

In addition to the technological constraints, the CI team encountered corporate constraints and resistance. The CI team's vision of the user's experience required significant development resources. The team looked not just at a new manual or help system but significant application features that required a large commitment from development in time, money, and people. Development resources were needed for building a Communicator, Watcher, Tipsheet Manager, and so forth.

Other constraints included cultural and organizational issues. The contextual design process was relatively new to the company, and acceptance by others within the company took time. The process required a new way of thinking and doing things, allowing individuals and groups to be involved in the process in new ways. For instance, individuals from the documentation department had never had such a

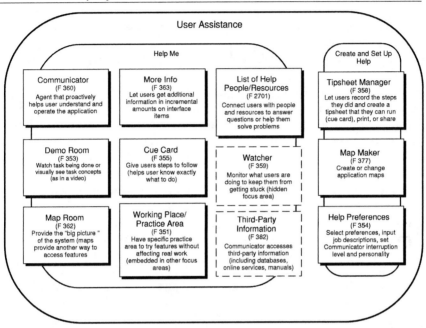

Figure 8.4. Formalized User Environment.

significant or early role in development process. Acceptance of new ideas and methods required artful use of interpersonal skills by the CI team, as they lobbied for support in the company.

Interestingly enough, some of the greatest resistance came from others in the documentation department (technical communicators and information designers) who had not been involved intimately in the process. Although the CI vision and model was never intended for wholesale implementation in one release, even its incremental implementation threatened the traditional way of doing things. Many writers and editors resisted becoming more involved in interface issues and "documentation" integral to the application, such as long prompts and error messages. The overall attitude of the documentation group was not progressive. Many who had resisted the implementation of online help also resisted the vision of user experience. Collectively, the documentation department had established itself as a producer of printed manuals and online help, and many felt threatened by the new skills and roles required by the CI vision.

Perhaps the major organizational constraint affecting the CI work resulted from a business event totally independent of development and the data: The company was sold. Although the new company ultimately endorsed the work of the CI team, new managers and personnel needed to be convinced of both the process and value of the contextual design approach. The organizational transition curtailed the implementation of the vision, at least in the short term.

Ultimately, some envisioned changes were implemented in the next release. Although the vision of the total user experience was not realized, several significant changes were made. The next product releases included a limited intelligent agent that assisted users in certain tasks. The documentation group did work more with redesigning error messages and implementing pop-ups to display critical information, both ideas grounded in the CI data. Additionally, elements of a Natural Language Interface were integrated, particularly in the Help system (building on the data that users want information in their terms and in language they use and understand).

Implications for practice and research

This chapter demonstrates the contextual design process applied to information and documentation and provides some of the results of the inquiry process. As an interpretive case study, it shares insights the team learned about users. The proprietary nature of some of the data and space limitations prevent detailing every aspect of the process or all of the results. However, the study generates important implications for practice and implications for further research.

One of the most significant insights gained by the team was the need to expand their definition of documentation and reassess its value. Too frequently, documentation is viewed as just a printed manual and an online help system. Documentation is not just about user education or user assistance; rather, it is an integral part of an application and users' perception of it and significantly impacts the entire user experience. As Mayhew (1992) observed,

> User documentation is part of the user interface. Often it is a user's first encounter with a system. Users may form impressions of the usability of a system by reading the manual or accessing the online help or tutorial. If the manual or tutorial is easy to use, users will assume the system will be easy to use. When customers are deciding whether or not to buy software packages by scanning through the manual, the quality of the manual can influence their decision. The quality of the user documentation thus determines in large part the perception of and the actual ease of learning and use of the system. (p. 534)

Although the CI team did not start out narrowly just to create a new manual or help system, their understanding and appreciation of users' experience nonetheless grew as they increasingly worked with customers. The team quickly expanded their vision of user assistance to system-wide implications that involved the whole user experience. They realized that the interface and documentation are integrally linked, that the way things are named and ordered is integral to users' experience, that menus are a first level of documentation, and that error messages and system prompts are significant points of system interaction where users learn a system or recover from system problems. The realization grew that what we traditionally view as documentation—printed manuals, online help, Web support, tutorials—is all part of the "interface": the bridge between the application and the user. The CI team realized, as Hackos and Redish (1998) claimed, that genuinely successful applications need to reflect the workflow of users, support various users' learning styles, and remain compatible to users' working environment and language (pp. 5-7).

Through the process of their work, the CI team found that the contextual design methodology provided a valuable method for gathering customer data in a systematic, usable way, that information about users and their work is best gathered through the process of observing and working with users in their own environment in the context of their own work. Contextual design has application for developing documentation and information not just developing systems and applications. Frequently, information designers and technical communicators view themselves as users advocates, with the mistaken notion that as nondevelopers who also work with an application designed by someone else, they are closer to the user. The CI team found that if they truly wanted to be user advocates, they needed to be out with users, and the contextual design process provided a method for finding out what users actually do, not just what they think they do or want.

Particularly with resistance coming from others within the documentation department, the CI team observed that it is difficult for many technical communicators to think strategically, to "get out of the box" and realize they can and should do more than just write or edit. As many have advocated, technical communicators need to be willing to learn new skills and be more involved in the whole development process (Bradford, 1991; Bresko, 1991; Fisher, 1999; Zimmerman, 2001). Such a shift requires technical communicators to learn new skills and be willing make changes in what they do and how they do it. If they are willing, the growth of information design can provide an opportunity for them to better position themselves as strategic players within organizations, to better understand and support how people interact with applications. As organizations move toward increasing customer satisfaction, technical communicators (or information developers) can significantly impact customers' experience through improving users' experience.

In addition to implications about the discipline and work of technical communicators, the CI study provides important insights about users that have implications for how we design documentation as well as applications. The research provides valuable insights about users:

- Users need and want to use their own terms. Language or terms not consistent from application to application causes problems.
- Because of language and terminology differences, indexes are a critical way for users to access assistance. Indexes provide the entry point for manuals and help systems.
- Users seek assistance close to them, often to a neighbor or pod-mate. The closer the assistance to the user, the more reliable and less costly.
- Users create their own documentation.
- Quicktips (long prompts or pop-ups) and immediate contextual information (including error messages) are very important.
- Users seldom have time for formal training and traditional tutorials are seldom used in the context of real.
- Users usually stick with the first way they have learned something. Changing learned behaviors and methods is difficult. They do not care if their methods are inefficient.
- Users get angry when new versions "trash" their knowledge base by "improving" the software. They resent improvements that destroy their way of doing things.

These insights have important implications for how we develop assistance for users. For instance, applications need ways to allow users to search for support using their own language. Natural language interfaces work toward this end, allowing users to type queries in their own words. Additionally, indexes (an important entry strategy) need to extensively cross-reference terms, building on feature and task synonyms. Also, visual indexes that show elements of the program or products of the application can identify key features for users. More strategies and techniques that allow users to point and identify features will further assist users in determining how they can best accomplish desired tasks.

Because users create personalized documentation according to varying contexts and needs, applications need to better assist users in creating, saving, and sharing the documentation they create. As systems become more aware of users' actions, they can track work and steps to create tipsheets or cue cards that become a formalized part of the users' experience. Increased awareness of users, their preferences and learning styles, can help provided assistance to be more personalized and appropriate for given contexts and needs.

The information gathered from the CI study and other customer-centered design methods has various application and potential outputs. Technical communicators, information designers, and system developers can use data from customers to modify existing tools, systems, and documents, or the information can be the basis for developing new applications, systems, and documentation as well as assist in developing new work practices. Such work, however, requires an awareness of and working within organizational constraints as well as awareness of the customer. The CI team realized that even with great ideas, they needed others in the organization to have understanding and "buy-in" along the way. Design teams need to assess what ideas can be immediately implemented and which ones can wait until later versions. For instance, although the CI team could not implement a complete system that tracked and communicated with users, they were able to develop a more extensive index for printed manuals and implement a limited Natural Language Interface in the Index/Search feature of their Help system that moved toward providing assistance in the user's own language.

In addition to applied implications for practice, the CI data has important implications for additional research. This data captured information about users at a specific time. Any customer-centered approach is iterative and requires continual work with the customer. As technological advances make possible the CI team's vision, the data must be validated through additional contextual inquiry and research. Technology will continue to allow new methods and ideas to alter the work and work practices of users. Innovative ways to best enhance a user's experience can be discovered and implemented through continual work with and research about users.

The CI team gathered information on how people learn and how they get unstuck, but additional user-centered study can provide insights on how users best perform tasks. Although some information was gathered tangentially about task performance, a focused study on the way users complete tasks would provide additional clues for enhancing a user's experience. Additional research can also target user learning styles and strategies with specific tasks. What type of information in what form promotes learning for what type of users in which contexts?

A continued thrust of user assistance has been to provide more user interaction with computers. The use of electronic assistance that gives timely and correct information continues to grow. The concepts of the Watcher and Communicator provide a vision for intelligent agents. Although initial research indicated users wanted an actual entity to simulate a person, the CI team's experience found that this often annoyed or distracted users. Further work needs to explore how users might best interact with systems. Are there ways that users can interact with systems directly, not anthropomorphized objects? What can we learn from failures of intelligent assistance, such as Microsoft's Bob—or its subsequent iteration, the Paper Clip (Smith, 1995)?

In a rapidly changing world users are in continual need to learn new tools, procedures, and methods, although they do not have time for formal learning. How can we facilitate learning? How can we apply what we know about learning and learning strategies to technical communication and information development? What constitutes a "safe" environment that encourages users to explore and apply learning?

These and other issues are central to the emerging discipline we call information design. As Alan Cooper (2001) observed, "the great promise of the information age is that computers help us do everything. The great tragedy of the information age is that computers obstruct everything we do" (p. 2) User-centered information design holds the promise of better bridging and mediating the gulf between humans and technology as the value and importance of information grows. Information design has the unique opportunity for helping us to understand better how humans interact with computers and other machines and for finding innovative ways to facilitate and assist that interaction. As businesses compete in an increasingly competitive global market, the appropriate design and delivery of information will grow more crucial, for as Porter (1995) claims, information will become a significant dimension that differentiates products and services.

This chapter shows a method for gathering data about customers and how to translate that data into actual information design. Such methods increase the likelihood of designing products and systems that genuinely meet customer needs. The user-centered design of information holds the promise of helping organizations improve the experience of their customers and users by facilitating and increasing product usability, usefulness, and value—a goal that benefits not only the bottom-line of organizations but the every-day work experience of individuals.

References

Beabes, M. A., & Flanders, A. (1995). Experiences with using contextual inquiry to design information. *Technical Communication, 42*(3), 409-420.

Beyers, H., & Holtzblatt, K.. (1998). *Contextual design: Defining customer-centered systems.* San Francisco: Morgan Kaufmann.

Bradford, D. B. (1991). The new role of technical communication. *Technical Communication, 32*(1), 13-15.

Brassard, M. (1989). *Memory jogger plus.* Methuen, MA: GOAL/QPC.

Bresko, L. L. (1991). The need for technical communicators on the software development team. *Technical Communication, 38*(2), 214-220.

Carroll, J. M. (1995). *Scenerio-based design: Envisioning work and technology in system development*. New York: Wiley.

Cooper, A. (2001). *The lost chapter of TIARTA (The Inmates are Running the Asylum)*. Usability Professionals' Association Tenth Annual Conference Proceedings, Lake Las Vegas, NV, section 1:1-4.

den Buurman, R. (1997). User-centered design of smart products. *Ergonomics, 40*(10), 1159-1169.

Fisher, J. (1999). The value of technical communicator's role in the development of information systems. *IEEE Transactions on Professional Communication, 42*(3), 145-154.

Grech, C. (1992). Computer documentation doesn't pass muster. *PC Computing, 5*(4), 212-214.

Hackos, J. T., & Redish, J. C. (1998). *User and task analysis for interface design*. New York: Wiley.

Horn, R. E. (2000). Information design: The emergence of a new profession. In R. Jacobson (Ed.), *Information Design* (pp. 15-35). Cambridge, MA: MIT Press.

Isensee, S., Pierce, E., & Righi, C. (2001). Then a miracle occurs: Translating data into design. *Usability Professionals' Association Tenth Annual Conference Proceedings*. Lake Las Vegas, NV. Section 4:1-33

Irvin, S. (1996). Documentation. *InfoWorld, 17*(46), 132.

Jacobson, R. (2000). Introduction: Why information design matters. In R. Jacobson (Ed.) *Information Design* (pp.1-10). Cambridge, MA: MIT Press.

Kawakita, J. (1982). *The original KJ method*. Tokyo: Kawakita Research Institute.

Keil, M., & E. Carmel. (1995) Customer-develop links in software development. *Communications of the ACM, 38*(5): 33-44.

Leonard, D., & Rayport, J. F. (1997). Spark innovation through empathic design. *Harvard Business Review*, November-December (pp. 102-113).

Mayhew, D. J. (1992). *Principles and guidelines in software user interface design*. Englewood Cliffs, NJ: Prentice-Hall.

Nielsen, J. (1993). *Usability engineering*. San Francisco: Morgan Kaufmann.

Norman, D. A. (1989). *The design of everyday things*. New York: Doubleday/Currency.

Norton, D., & Hanson, L.. (2000). The e-commerce blueprint: Creating online brand experiences. *Design Management Journal*, Fall, 25-35.

Parasuraman, A. (1997). Reflections on gaining competitive advantage through customer value. *Journal of the Academy of Marketing Science, 25*(2), 154-161.

Passini, R. (2000). Sign-posting information design. In R. Jacobson (Ed.) *Information Design*, (pp. 83-98)., Cambridge, MA: MIT Press.

Pine, B. J., & Gilmore, J. H.. (1999). *The experience economy: Work is theater and every business a stage*. Boston: Harvard Business School Press.

Porter, M. E. (1985). *Competitive advantage: Creating and sustaining superior performance.* New York: The Free Press.

Raven, M.E., & Flanders, A. (1996). Using contextual inquiry to learn about your audiences. **The Journal of Computer Documentation, 20*(1), 1-13.

Rettig, M. (1991). Nobody reads documentation. *Communications of the ACM, 34*(7) 19-24.

Rust, R. T. & Zahorik, A.J. (1993). Customer satisfaction, customer retention, and market share. *Journal of Retailing, 69*(2), 193-215.

Schuler, D. & Namioka, A. (1993). *Participatory Design: Principles and Practices.* Hillsdale, NJ: Erlbaum.

Smith, R. M. (1995). Microsoft Bob to have little steam, analysts say—Ask if 'social interfaces' are the best solution. *Computer Retail Week*, April 3, 37

Wood, L. E. (1998). *User interface design: Bridging the gap from user requirements to design.* Boca Raton, FL: CRC Press.

Zimmerman, M. (2001). Technical communication in an altered technology landscape: What might be. *Technical Communication, 48*(2), 200-205.

9

Dynamic Usability: Designing Usefulness Into Systems for Complex Tasks

Barbara Mirel
University of Michigan

Complex tasks and problem solving are different in kind not just degree from well-structured tasks. They require unique designs for software support. Stressing the primacy of usefulness in designs for complex work, I propose a framework — Bounded Interactivity for Complex Inquiry (BICI) — to guide usability specialists, information designers, and other members of software teams in building usefulness into programs for complex problem solving. BICI emphasizes the interaction between contextual conditions, constraints, and actions in order to guide software teams in creating useful designs centered on the effects of these interactions. I exemplify this conceptual framework by applying it to a sample design situation.

Demands of complex tasks

Increasingly, software developers are creating programs to support complex problem solving. One such application is root cause analysis in medicine, the process of finding the causes of adverse events like medication errors or faulty surgical procedures (Autros™, Root Cause Analyst™). Reducing such errors is a leading national agenda item in health care today (Kohn, Corrigan, & Donaldson, 1999). In tracing adverse events to their sources and devising remedies, risk managers

233

in medicine are beginning to take a systems view. They realize that long-lasting reductions in risk can only come from analyzing relationships among systems of work, technology, environment, and communications, not from blaming and penalizing individuals. A systems approach, for example, may reveal that the nurses and pharmacists involved in an overdose event were overloaded with work, which, in turn, was linked to their having to use a new automated system for medication that their hospital implemented before allocating adequate resources for infrastructure upgrades and without restructuring staffing as needed.

Analyzing root causes for medical errors is just one of many complex, real-world tasks and problems that software developers today are trying to support. Unfortunately, in the software industry, strategies, methods, and implementations of software for complex tasks and problems are still in a fledgling stage. Software and user interface designers who traditionally have been concerned with this issue—those who create decision support systems or complex information retrieval systems—are still struggling to make systems "adaptable enough." Being adaptable enough means that a system takes on the role of a partner in users' job-related, open-ended inquiries. For this adaptability, software designs need to "lean on" the domain expertise, professional practices and conventions, situational knowledge, and decision-making criteria that *users* bring to their program interactions (Agre, 1997).

Software that leans on users' problem-solving practices and expertise differs from typical software that supports users' tasks. Typically, software provides prepackaged procedural solutions and easy and accessible operations for executing them. By contrast, designs that lean on users' expertise structure virtual work spaces in ways that give users many options and degrees of freedom to choose the actions most appropriate for their purpose and context. They visibly structure the boundaries and conditions of a certain type of work-related problem in ways that tap into users' expertise so that they recognize the type and its associated patterns of inquiry strategies and actions. Adaptable software also leans on users' worlds by giving them control over the approaches that they may take to multidimensional, multi-pronged inquiries. To achieve effectively the goals of complex problem solving, users characteristically need to be able to take their own approaches to arranging information and other resources for multiple points of view. They also need to be able to filter and scale information as needed and to improvise as they explore. For open-ended and emergent work, adaptable interfaces need to be dynamic to facilitate flexible uses (Hollan, Hutchins, & Kirch, 2000).

In the literature on usability, designing for the adaptability that is integral to complex work has received only recently mainstream visibility (*Human Factors*, 2000; *ACM Transactions on Computer-Human*

Interaction, 2000; Carroll, 2001). In industry such design practices have been slow to catch on, especially in companies with technology-driven rather than market-centered structures and processes. Usability specialists and other information designers have a critical role to play in designing useful programs for complex tasks. They can do so by informing this software with perspectives and designs attuned to the demands of dynamic work.

To discover interface designs that are critical for supporting complex problems solving, I have studied over 100 problem solvers in 30 different contexts of work as they conducted various types of complex inquiries during their daily jobs. In this chapter, I draw on findings from my study to argue that the most important criteria for supporting complex work is to *design for usefulness*, a different focus from designing for ease of use and access. I then propose a framework — Bounded Interactivity for Complex Inquiry (BICI) — to guide usability specialists, information designers, and other members of a software development team in building usefulness into programs for complex problem solving. Representing users' complex problem solving in context, BICI emphasizes the interaction between conditions, constraints, and actions, and the need to design programs and interfaces in ways that support the effects of these interactions. After explaining this conceptual framework, I exemplify how it can be applied by examining a sample design situation. This situation focuses on design strategies for one of the most common problem-solving needs experienced by users in my study, namely the need, in as few keystrokes as possible, to carry out and see results from the methods of analysis that are central to their domain and profession's way of inquiring into a problem. I conclude by addressing the implications of designing for usefulness.

Usefulness for complex problem solving

In designing software support for dynamic work, usability specialists and information designers need to assure first and foremost that the software is useful. Useful software helps people do their work. By contrast, easy to use software helps them work the program (Ziech, 2000). Before being concerned about ease of use, designers and usability specialists need to be sure that a program embodies the right model of users' goals and work in the first place. Otherwise—as is the case unfortunately with many programs for complex problem solving—software teams may create programs that are accessible and easy to operate but pragmatically meaningless. This concentration on designing for usefulness needs to be brought into front-end discussions and decisions about scope, architecture, and features as well as interface designs.

Vincente (1999), a long time proponent of innovative interface designs for complex problem solving, discusses this need for software design teams to focus on usefulness more than on easy to operate controls for preset task procedures and navigation. For complex problems and decisions, he stresses, designers must get the work model right in the first place. He argued:

> The low hanging fruit [of automating rule-driven activities] has already been picked. What remains is a much more challenging problem—how to support human intellectual activities that we are not capable of automating fully. Such activities usually require workers to engage in open-ended and creative, discretionary decision-making (which is largely why these activities cannot be fully automated). However, designers have not been able to design effective information support for these intellectual tasks... [T]he trouble with computers is not just that they are not very easy to use but also that they generally do not provide people with the functionality that is required to get their jobs done efficiently. (p. 26)

Assuring useful software for complexity is a challenge because it involves knowing and designing for the many ways in which people relate information and arrange other material resources to resolve open-ended problems. These problems arise from and change with contextual circumstances and goals. Complex tasks such as root cause analysis depend on and vary with events and contexts; the same task is rarely if ever performed the same way twice. The means of performance are not known at the outset. Approaches to complex work emerge and become better specified as people explore conditional factors relevant to the issue or problem at hand. Knowing what to do next requires coordinating, arranging, and relating numerous factors. Typically, several renditions of acceptable arrangements exist. Similarly, the rules, heuristic strategies, and trials-and-error that people apply lead to processes that allow for several outcomes (Vincente, 2000). Many standard office applications such as spreadsheet programs may permit complex work as well as rote data entry but their designs do not explicitly aim to support these defining qualities of complexity.

For example, if a marketing analyst for a coffee manufacturer is inquiring into whether a new espresso product is likely to succeed in this specialized market, she needs to view, process, and interact with a wide range of multiscaled data. To figure out what it will take to break into and become competitive in the high-end espresso market, the analyst will examine as many markets, espresso products, and attributes of products as she deems relevant to her company's goals, and as many as her technical tools and cognitive capacity enable her to analyze. Looking at these products, she will move back and forth in scale between big picture and detailed ("drill-down") views. She will assess how espresso has fared over past and current quarters in different channels of distribution, regions, markets, and stores, and impose on the data her own knowledge of seasonal effects and unexpected market conditions. For different brands and products—including variations in product by attributes such as size, packaging, and flavor—she might analyze 20 factors or more, including dollar sales, volume sales, market share, promotions, percentage of households buying, customer demographics and segmentation. She will arrange and rearrange the data to find trends, correlations, and two-and three-way causal relationships; she will filter data, bring back part of them, and compare different views. Each time she will get a different perspective on the lay of the land in the "espresso world." Each path, tangent, and backtracking move will help her to clarify her problem, her goal, and ultimately her strategic and tactical decisions.

People's actual approaches to complex tasks and problems such as this marketing analyst's are contextually conditioned, emergent, opportunistic, and contingent. Therefore, complex work cannot be formalized into formulaic, rule-driven, context-free procedures. Programs will not be optimally useful for complexity when they precompute plans and presuppose and map rules and steps onto program features and commands (Wright, Fields, & Harrison, 2000). Such programs offer problem solvers preset routes through stable territory when what users need is help in exploring territory that is uncharted for their specific purposes. In this latter situation they need to understand the structure of the territory and manipulate it as needed to find both a valid route and destination. Program and interface design for this situation requires new conceptual frameworks for thinking about task support. These frameworks, in turn, need to inspire new ways of designing program displays so that they provide adequate and adaptable maps for complex problem solving.

A framework for designing for usefulness

To support complex tasks and problem solving, programs need to provide people with the adaptability and flexibility that dynamic inquiries demand and to give them the ability to control and coordinate the core aspects of inquiry that are fruitful only when users have this freedom. The coffee analyst mentioned earlier would be hamstrung in her decisions, for example, if she could not backtrack at will and if, in her back and forth progress, she could not coordinate her emerging discoveries and their implications to fit her purposes. Open-ended explorations and inquiries into unanticipated events cannot be adequately supported with rigid, albeit well laid-out, procedures—represented through program functions (for instance, save, sort, or search) and executed through clearly labeled and accessible interface controls. Such a procedural orientation in design does not offer the degrees of freedom that users need in order to construct and continuously shape inquiry approaches to contextually based purposes and emerging conditions (Rasmussen, Pejterson, & Goodstein, 1994). Nor do predominantly procedural orientations provide users with an adequate map of the "complex problem territory." More than procedural sequences for preset routes, problem solvers need to relate the "criss-crossed landscape" of knowledge and actions in their given problem spaces to their goals, which are often vague and evolving. They need flexibility so that they may clarify and revise initially uncertain and broad goals and shape the course of their inquiry to emergent discoveries. Finally, a procedural emphasis in interface support is insufficient for complex problem solving because, even with "intelligent agents," it cannot possibly anticipate or capture in preset steps all possible approaches (Agre, 1999). Complex work, by definition, is situated and has an element of unpredictability. In complex inquiries, "knowledge is in the connections" or relationships, not in discrete components of the task or of its rule-based processes (Bereiter, 1991, p. 15).

Vincente and Rasmussen (1992) are strong critics of typical procedural approaches and the predominating focus on ease of use and access. Concentrating on complex troubleshooting and fault management in process control systems, they stress the inadequacy of these approaches for the demands of complex problem solving. They argued:

> [M]uch of the work that has been done on interface design for complex systems has focused on these types of ergonomic issues [clearly labeled controls and displays] rather than on semantic issues…[E]fforts to identify the information set needed by operators to deal with off-normal events have tended to adopt a common approach. A set of events is selected and then the information needed to diagnose each event sequence is determined. But this type of procedure for determining what information should be included in an interface cannot, by definition, cope with unanticipated events. Clearly an alternative approach is required. (p. 590)

The alternative has to shift from this dominant focus on building support and information displays around task models that consist of procedural sequences. A procedural emphasis may work for many formulaic, rule-driven tasks but not for dynamic, emergent work. An alternative that many specialists in cognitive systems engineering and activity systems increasingly envision for designing support for complex work is a structural orientation that calls forth the constraints or boundaries of users' work—the "shape" of their work—and the conditions under which the uses of certain information, actions, objects, and relationships have meaning (Agre, 1997; Dillon & Vaughan, 1997; Rasmussen et al., 1994; Vincente, 1999).

As seen in the coffee analyst example, when problem solvers conduct open-ended and ill-structured inquiries, they draw cues about what to do from the conditions and constraints of their situation. If the coffee analyst knows that sales for coffee were unseasonably low during Christmas, she may decide to recalculate those quarterly figures speculatively. She will adjust them so that they reflect more "normal" cyclical circumstances for this season, giving her the opportunity to judge the market under more normal trends, as well. Conditions such as seasonal aberrance shape the analyst's range of available and relevant actions. By relating conditions, constraints, and possible actions to one another, such problem solvers choose the moves and strategies that are best for their purposes and circumstances. One set of conditions—seasonal aberrance—prompts this analyst to spend time manipulating, deriving new values, and comparing data. But other sets of conditions interact with this market condition, as well, to shape, redirect, or confound an individual's choice of actions. For example, technological conditions within the data analysis program may make it difficult or impossible for the analyst to derive new values and make comparisons within the program. In such a case, her trajectory of problem solving either grows far more complicated as she seeks another means of transforming and comparing data or it becomes partial and prone to inaccuracies as she decides to skip this phase of analysis for want of program functionality. In complex problem solving, each choice may reshape the focus, conditions, and boundaries of a workspace and suggest a need for revising goals, setting in motion a perpetual process of dynamic and emergent investigation. If software support does not adaptively support the content and structures of users' evolving analysis, it may undermine the validity of their conclusions.

A framework of bounded interactivity for complex problem solving (BICI)

As Rasmussen et al. (1994) and Vincente (1999) argued, in computer-supported complex problem solving, one of the most important objectives of software design is to cue users to the relationship between their goals and the means of performance. These relationships need to be cued for many interconnected levels of activity, all of which feed into the whole of a user's work space. These levels include cognitive, social, and technological dimensions of work. With the need for this cueing in mind, Vincente and Rasmussen (1992) proposed a conceptual framework called Ecological Interface Design (EID). Their EID model centers on representing the conditions and constraints that occur in various stratified levels of users' actual work space and make apparent the relationship between the conditions and constraints occurring in these levels and the purposes and actions that problem solvers carry out within them. The levels are interactive. They reciprocally shape the actions that problem solvers may pursue and are bound together by the goals of the investigation. The framework that I propose next is an adaptation of EID.

When Vincente and Rasmussen originally created the EID model, they intended it to meet the specific demands of monitoring and troubleshooting large, complicated process-control systems. The work space levels, trajectories of work, and possible actions represented in EID target problems that are grounded in and solved within the workings of these physical systems. The machinery is the work space. For this work space, Vincente and Rasmussen propose the following levels as the constituent aspects of work that condition and constrain problem solving behaviors:

- The purposes of the work system (the process control system) and the constraints that occur when it is coupled with users' work environment (functional purpose).
- The causal structure of work processes and priorities, their flow and accumulation of information, people, and value (abstract function).
- The basic functions that a work environment (the plant) is supposed to achieve (general function).
- The component traits, physical processes, and interconnections—relevant for knowing what to repair (physical function).
- The spatial location of components—relevant for navigating the system (physical form).

The creators of EID stress that it is generalizable, fit for analyzing work and designing systems in a variety of domains and work spaces. It has proved successful when applied to designing software for complex fault management and troubleshooting tasks in process control systems (Burns, 2000; Vincente, 1999). Rasmussen et al. (1994) also have applied it to the design of an information system for conducting bibliographic searches in library catalogs, with the caveat that these tasks are not as complex as troubleshooting unexpected breakdowns are. Despite these researchers' claims, the generalizability of the model may be more limited than they acknowledge, especially in regard to its ability to transfer to all work spaces and because of its high level of abstraction.

In regard to other work spaces, EID and its constructs do seem to transfer well to the problem-solving activities that I have studied in work spaces similarly comprising complicated, automated, continuous process systems, namely telecommunication as well as data networks. However, other work spaces that I have studied are not neatly captured by the EID model. In these other work contexts, tapping into and exploiting the logic and workings of a technical system—be it a continuous process or information retrieval system—do not constitute the means *and* ends of problem solving. Problem solving in these other work spaces may be supported by software for data analysis or cooperative work but making the software work is not the end or purpose of problem solving. Rather the goal is to answer a pressing job-related question. For example, I observed hospital nurses as they worked in patient care work spaces to administer medications and solve dosage and diagnosis problems. I shadowed business analysts as they worked in corporate work spaces to figure out how to expand market share and how to reallocate resources. Their work spaces comprise people, organizational standards, material resources, and a universe of data on commerce, marketing, and finance

These instances of problem solving took place in what Rasmussen et al. (1994) call "flexible production" work spaces, that is, in the work world of producing goods and services. In analyzing data to solve complex business problems, users have to share control in problem solving and decision making with the program that supports their inquiries. But neither the workings of the program nor a mastery of its search-and-retrieve mechanisms is the object of their work. "Flexible production" work spaces are "loosely coupled systems governed by actors' intentions ... in which the coordination and control of activities depend on the communication of company-institutional objectives ...[and] many degrees of freedom remain to be resolved by situational and subjective criteria by the staff... creating an intentional structure...more complex and dynamic than for the industrial process [work] systems" (Rasmussen et al., 1994, p. 52). For flexible production work spaces, the EID model needs to be modified. Its last two levels, for example, on the interconnections and locations of the components of physical or virtual systems do not map readily to human and social work spaces.

Based on my field observations of people solving complex problems in telecommunications, intranet management, hospital nursing, product marketing and sales, and human resource management, I have attempted to adapt the EID levels and their effects on problem-solving behavior into a more generalizable model. I also have tried to make it more accessible. As powerful and comprehensive as the EID conceptual rendering of complex problem solving is, it is also intellectually demanding. Its constructs and terminology are highly abstract. My aim is to provide a framework that is easier for designers to grasp and directly apply to their commercial software projects.

In my Bounded Interactivity for Complex Inquiry framework, I have sought to maintain the spirit of the EID constructs and their strong "family resemblance" to components used by many other models that similarly strive to capture how people conduct complex activities in context. This family of models includes activity theory, distributed cognition, situated cognition, cognitive flexibility theory, models of complex and emergent systems, models of user activity in information usage, and models for representing findings from contextual inquiries (Barnard & May, 1999; Beyer and Holtzblatt, 1998; Carroll, 2000; Cole & Engestrom, 1993; Dillon, 1994; Hollan et al., 2000; Holland, 1998; Hutchins, 1995; Kulak & Guiney, 2000; Marchionini, 1995; Nardi, 1997; Spiro & Jehng, 1993; Suchman, 1997). Beyer and Holtzblatt (1998), for example, suggest five frameworks for representing contextual inquiry findings that, similarly to EID and other family models, strive to capture functional and structural relationships among work processes, social roles, environmental dynamics, and physical arrangements. These researchers' representations take the form of workflows, task sequence and goal diagrams, models of artifact use, cultural dynamics models, and physical layouts of the work environment. Similar elements and interactions are captured by Cole and Engestrom (1993) in their visual rendition of distributed cognition (figure 9.1).

In addition to striving to stay true in BICI to the similar constructs used by this family of models to represent the complex dynamics of goal-based activity in a work system, I draw on Vincente (1999) and Rasmussen et al.'s (1994) later work. They expand on EID, tying problem solving behaviors to task requirements—task situations, cognition, and mental strategies—and to problem solvers' roles and resource profiles—role allocations in work environments and tasks, management style and culture, and subjective preferences and resources.

Similar to Vincente and Rasmussen's approach, I emphasize the interactions among conditions, constraints, and actions. Representations such as those proposed for contextual inquiry findings or for distributed cognition are comprehensive but are not structured in ways that highlight the mutually shaping effects of actions, conditions, and constraints. Rather these representations are composed and visualized in ways that foster a "reading" that highlights the actions that go on within and between components. An emphasis on action is not the most appropriate perspective for designing useful support for the emergent and serendipitous explorations that characterize complex problem solving. A high-

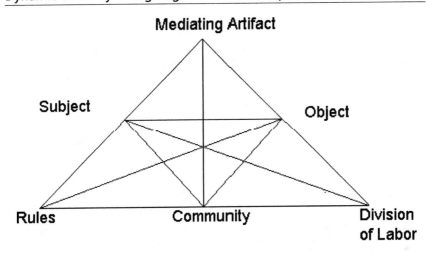

Figure 9.1. Elements of a work system over which cognition is distributed (From Cole & Engestrom, 1993).

lighting of actions too readily feeds into—intentionally or not—the common tendency to break work down to component actions and support it through procedural designs. As discussed earlier, an alternate to procedural design is needed for complex tasks. This alternate support has to stress relationships—the interactive effects of conditions, constraints and actions within and across the interconnected levels of the activity or work system. The advantage of Vincente and Rasmussen's EID model is that it highlights these interactions and effects. The problems with EID, as mentioned before, are that it may be too specific to the demands of process control system work spaces and that it is too abstract for designers to apply if they are not theoretically inclined.

My proposed Bounded Interactivity for Complex Inquiry framework generalizes and makes more concrete the EID levels, conditions, constraints, and actions. BICI consists of four levels, each with its own conditions and constraints that shape and direct the path of people's problem solving behaviors. These levels are condensed from the EID levels and are less abstract. They embody the same manifold contexts and contextual conditions as EID and other contextual models. The volume of these contexts and conditions may seem too much to account for in analyzing and designing for users' tasks. Therefore, in applying BICI to design, usability specialists and information designers need to focus selectively on what they observe to be the most salient interactive factors within and across levels, given the goals and requirements of their software project. Table 9.1 explains the levels.

Examples of conditions for each level include:

Problem—For a network troubleshooting problem, one condition may be a problem that occurs intermittently without much apparent rhyme or reason. Another condition may be a regularly recurring problem, far more easily diagnosed and solved. For each of these conditions, problem solvers look at different data and use different methods.

Work domain—A social and physical condition may be collaboration in troubleshooting involving different job roles (a database administrator and an application specialist) conducted remotely due to the physical location of each collaborator. An organizational condition may be standards for escalating and handing off certain levels of troubleshooting without further inquiry; and a professional condition may be the conventional use of a process-of-elimination method for troubleshooting, ordered in a certain way. These intertwined conditions make some but not other relevant actions possible.

Subjective—A subjective condition may be a person's cognitive capacity to process data from no more than 5 different tools and display modes at once while troubleshooting a breakdown.

Technology—A technology condition may be the capacity to store large volumes of data so that problem solvers have data available for comprehensive analyses.

Technology   artifacts—Another condition may be that a company's infrastructure includes the use of an emulator such as Citrix. In Citrix environments, network "sniffers" are unable to read any activity behind the Citrix "wall," thereby limiting the data available for troubleshooting.

BICI in Relation to Usefulness

Since getting the conditions and constraints right for various types of complex problem solving is crucial to designing for usefulness, it is important to supplement table 9.1 with further definitions and examples of what I mean by conditions and how they constrain and shape problem solvers' behaviors. The conditions under which users' actions and interactions have meaning are the occurrences in time and place that direct users toward certain subgoals, paths, and ways of thinking for a given type of work or problem. For example, in nursing in the work space of patient care, a nurse's activities are constrained and shaped by the type of problem that the nurse addresses (at the problem level of the

Table 9.1. Levels and conditions of a workspace.

Level	Conditions	Examples of conditions
The problem situation	Type of problem and goal Trigger event for inquiry Patterns of inquiry	In the retail world of category management, one type of problem is "evaluating product performance against a baseline." It involves a comparison pattern of inquiry. For example, users take the actions needed to compare the product's actual market share over the past four quarters to its fair market share (the baseline). A problem type and its associated patterns of inquiry shape the data that analysts examine and the methods that they use.
The work domain	Social structures, processes, relations Culture Organizational policies, standards Organizational economies Organizational politics Professional practices, conventions Physical arrangements External environment	A bureaucratic social structure will constrain a problem solver's scope, level, and methods of inquiry as follows: A hierarchical structure strictly divides and centrally controls the labor of problem solving. A low-level analyst, for instance, finds only the facts about market performance and passes these findings to his superior. The superior draws interpretations and, in turn, hands off the inquiry to the next higher level where a decision is made. Software support for hierarchical social conditions vary dramatically from the support needed for an alternate condition of cross-functional collaboration in flat social structures.

Table 9.1 (cont'd). Levels and conditions of a workspace.

Level	Conditions	Examples of conditions
Subjective factors	Cognition Perception Individual preference Motivation and morale Prior knowledge and experience	An individual's cognitive condition of domain expertise and knowledge lead her to examine different factors and relationships from domain novices. In terms of motivation, a motivated problem solver is more likely than an unmotivated one to avoid make-do solutions or work-arounds.
Technology	Software features, functions, algorithms, built-in intelligence Architecture Data and database structures Data integration Information technology infrastructure	Programs that have the condition of little or no integration of data across programs or databases constrain problem solvers in the relationships that they may draw to answer their question.

work space). Of the many problem types that occur in patient care, one is administering medications. Within this problem type various conditions may arise—for instance, the condition of needing to assess a patient to determine dosage or of needing to verify a medication record and correct faulty entries. These two distinct conditions of medication administration call forth different possible actions and interactions with coworkers, programs, information, and other objects.

Within a given condition different triggers prompt problem solving that are not easily predicted and formalized into flowchart sequences. Each trigger may lead a nurse to look into the same condition in different ways. For example, a nurse in a detoxification unit may be triggered to assess a patient's condition and determine dosage by the routine requirement to taper every patient's detox medication regime at each scheduled medication pass. For another detox patient, the same nurse may be triggered by a sudden complication in the patient's condition and need to take a different course of action regarding dosage. Because of the complication, the nurse has to reconsider the routine detox drug regimen altogether and consult with the attending physician immediately to discuss the possibility of altering the patient's drug treatment. These two triggers prompt the nurse toward the same high-level activity, namely assessing patients' conditions and determining medication needs. But, in each case, because of trigger conditions, the nurse looks at different variables, takes different sequences of actions, and, in the latter case, brings other people and consultations into the problem-solving task. Trigger events shape problem solvers' strategies and tactics.

Designing for usefulness has to account for differences in problem solvers' approaches. Paradoxically, one way that designers can account for difference is by representing the *regularities* that underlie the surface differences and by depending on these represented "deep structure" regularities to "partner" with users' domain expertise (Pentland & Rueter, 1994). When evoked, these regularities will cue problem solvers about the distinctive surface level actions that are relevant to their situated conditions. For example, in the two dosage cases noted earlier similar structures occur for the process of assessing patients' conditions; correspondingly, relationships between vital signs and other indicators of health have similar meanings even if the exact paths taken in making these assessments vary. Programs are useful if they reveal and give users access to these underlying regularities. From the regularities, users are cued to the appropriate and available inquiry patterns. They are then able to apply their expertise to choose and construct the unique surface path that they need to take.

Designers need to focus on and represent interactions between conditions, constraints, and possible actions within and across levels because programs are not useful when they push users too close to the constraints associated with any given condition or set of interdependent conditions. Programs may push users to their limit when their designs (at the technology level of a work space) misgauge and misrepresent the constraints that users actually experience in the other levels of their work space. In terms of subjective constraints, for example, users have certain cognitive limits for recall and for multi-step calculations. Programs misgauge the conditions and constraints of users' subjective level of the work space when they demand a memory load or mental math that outstrips users' cognitive capacities. Programs misgauge work domain constraints—such as the time pressures that occur in the constant emergency pace of a hospital Intensive Care Unit—when the actions and paths that they allow for users' inquiry patterns take too long for the real-time demands of a user's job in-context. And they misgauge professional, technological, and social constraints if they set too narrow a scope and if they lack the functionality required to conduct problem solving in ways that stay true to users' sense of professional identity and social role.

One example of this misgauging of users' professional constraints in their everyday work lives is a program for administering medications safely that sets its scope narrowly on patient safety rather than broadly on patient care. Pragmatically, in nurses' professional work lives, medication administration involves more than checking for safety. When programs facilitate safety alone, functions and features may help nurses enter drug-giving information and assure the safe matching of the right drug to the right patient during medication passes, but they leave out other medication activities that from nurses' perspectives are intrinsic to the practices of their profession and ethos of care giving. In a medication administration program that I observed dozens of hospital nurses using, the software did not offer easy and reliable ways for nurses to comprehensively and electronically communicate to the next shift about unpredictable patient reactions. Nurses were not willing to forego the practice of communicating across shifts. It was essential to their professional identity and their socially shared schema for medication administration. But to carry out this practice in their software supported work, they had to move very close to the boundaries of the documenting and reporting features that designers had modeled and built into the medication administration program. These features were created specifically for safety issues such as recording dosage amounts and entering predefined comments about drug effectiveness. They were not designed to facilitate the kinds of interchanges between nurses that conventionally occur across shifts. At the boundaries of these features—the built-in (misrepresented) models of professional practice—nurses searched for some set of actions that might help them communicate across shifts, even though such actions were not apparent.

In this situation, users' actual professional conditions, inquiry patterns, and associated boundaries of work did not match the program model of professional practice that was built into its features. The result was that, in search of adequate support, nurses were pushed to the edges of what to them were acceptable practices for documenting and reporting and they lost control over doing their work in the ways that they needed and expected to do it. They sought to adjust program-defined conditions and boundaries to match their pragmatic realities—or at least to make them elastic enough to accommodate these realities. To do so they devised inventive program workarounds, time-consuming contortions, and cobbled together solutions. Once they had to resort to this interactivity, usefulness took a nosedive. It was undercut by decisions that the software project team made early in development, decisions that led designs to be discordant with the conditions of work and patterns of inquiry that users pragmatically experienced.

The medication administration project team had limited the program functionality to operations that assured safety, not collaborative medication administration processes. In doing so, they introduced technological conditions into nurses' work that kept nurses from assuming the full range of professional roles that held meaning in their hospital environments. This misfit between program functions and users' roles and work practices, in part, was the result of an even earlier decision about program scope. The chosen scope of medication safety rather than patient care did not match nurses' notion of the whole of their medication administration work. The scope did not account for the full range of nurses' problem types and inquiry patterns at the work domain and problem levels of their medication work space.

In a taxonomy of user errors that Sutcliffe, Ryan, Doubleday, and Springett (2000) proposed, they trace 3 out of their 11 categories of errors to these mismatches between functionality (the technology level of a work space) and users' actual practices (the work domain level). Either (a) "the system model of the task does not contain a function for the user's goal," (b) the system has the needed functionality but "support for the user's task could be improved," or (c) "the system model of the task partially achieves the user's goal but does not exactly match the user's expectations" (Sutcliffe et al., 2000, p. 45). The medication administration program reflected the latter of these three shortcomings. This example underscores that partial support is not good enough. To assure usefulness, information designers and usability specialists have to capture how users solve problems in-context, discover the core areas in which they need to be free to choose and control their open-ended pursuits, and design functionality and interfaces for them by stressing inquiry regularities and compatibilities between multilevel conditions and constraints.

Designing useful interfaces for analysis-in-a-keystroke

As the previous example reveals, it is crucial for software teams to acknowledge that the responsibility for usefulness cannot solely be delegated to user interface design. Usefulness is a whole-product quality. As seen in the medication administration example, it requires a program scope that accommodates users' notions of the whole of their work. It also requires adaptive functionality and architecture, often necessitating the use of advance technologies. For example, interactive information visualizations, intelligent agents, and integrative architectures for reading, writing, and merging data across several sources are powerful in supporting complex problem solving (Card, Mackinlay, & Schneiderman., 1999; Horvitz, 1999: *IEEE Computer Graphics and Applications Special Issue*, 1997; Lieberman, 1995; Maes & Shneiderman, 1997; Ware, 2000). Specialists in adaptive computing and intelligent agents have devised probabilistic models and inference systems that can "reason in uncertainty" (Horvitz, 1999). But, at present, this intelligence is applied mostly to well-structured, not complex, user tasks. In the design strategy that I propose next, the example assumes interactive information visualizations. A detailed discussion of technological advances, however, is outside the scope of this essay. But it is nonetheless crucial to unite technological advances and flexible interface designs to achieve usefulness. Designing for usefulness has to be a whole development team effort.

With this qualification in mind, I exemplify one design situation that shows the ways in which the BICI framework may be applied. This situation focuses on the need to support the analytical structures and methods that users devise for particular problems based on such social conditions as conventions of their professions and domains. Problem solvers "impose" these structures and methods on large amounts of multi-scaled and multidimensional data in order to make sense of the data for their specific purposes. In my study, problem solvers expected readily available and quick support for methods that were central to the ways their profession organized and conducted complex inquiries for specific problems. Without this support, they were often unable to complete their problem solving efficiently, accurately, and thoroughly.

Background on the design challenge for analysis-in-a-keystroke

Program and interface designers face a considerable challenge in having to anticipate and design for the methods of analysis that problem solvers in particular roles and domains must be able to carry out to resolve certain types of complex problems in their fields. Toward this end, functionality and interface design are intrinsically interwoven, especially when software presents data in dynamically linked, interactive graphics (figure 9.2). Such visualizations are increasingly being built into programs for complex problem solving. Interactive graphics visually display data of interest and allow users to directly interact with the data through the graphics. Directly manipulating graphics, users select and filter data, drill down, and discover relationships by strategically encoding dimensions into color, size, and other perceptual cues.

To be analytically useful, interactive data visualizations have to be designed to allow users to employ and see the results of the analytical

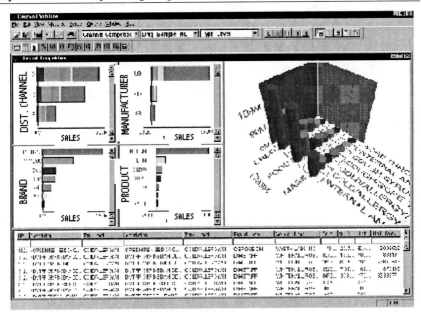

Figure 9.2. Multiple perspectives on "What items are doing well in sales?" On the left, the bar graphs present sales data for (1) each channel of distribution (mass market, grocery, and drug stores); (2) each manufacturer; (3) each brand; and (4) each product. On the right, for an overall quick view of the best-selling products by channel, the heights of the 3D bars show the sales amounts for each product in each channel. On the bottom, the table of textual data gives additional details for each product, including attributes such as package size and flavor and other measures such as expenses and market share. Analysts will want to look at the top selling brands (the longest bars in the Brand bar chart), select them, and find out their combined market share. The present display does not provide this information on combined market share.

methods relevant to the lines of reasoning in their particular area of specialty for a given type of problem. These lines of reasoning are not generic. They are social and contextual. They are the shared regularities by which people affiliated with a certain profession, domain, and role socially organize their work (Dillon, 1994; Pentland & Rueter, 1994). For category management analysts with their marketing problems, for example, these methods include setting a baseline for successful product performance and comparing their own and competitors' products against it. In my study, methods of inquiry were shared when people held their profession, job role, and domain of specialization in common regardless of whether they worked in the same company or office building.

Information visualization specialists have long recognized the need to categorize methods of analysis and design for them. Typically, they classify methods at a high, generic level and logically rather than pragmatically link them to graphics and functionality. Generic classifications include, for example, comparison, correlation, association, causal relation, conditional relation, clustering, distribution, frequency, ranking, trends, probability, what-if projection, and aberration (Bertin, 1999; Cleveland, 1993; Marchionini, 1995; Morse & Lewis, 2000; Senay & Ignatius, 1990; Shneiderman, 1996; Tukey, 1977; Tweedie, 1997). Links between methods and graphics include analysis of variance in box plots and correlations in scatterplots.

As helpful as such suggestions may be for an initial understanding of graphical support for distinct analytical methods, they deal largely with high-level rules of thumb drawn from statistics and principles for the graphic display of quantitative data and other structured information without reference to context. Based on this high-level orientation, it is not uncommon for software designers to attempt to build functionality for all the methods that they can imagine users might perform in the course of their various inquiries. The problem with this strategy is that it is neither cost efficient, feasible nor desirable from a usefulness point of view.

From the vantage point of usefulness, problem solvers do not simply want analysis functions to exist in their programs. They want to conduct specific analyses relevant to their professional lines of reasoning, and they want access to these methods at the point of need with little effort. For this support, guiding principles grounded in the context of work are needed to complement high-level, generic rules of thumb. Software for complex tasks is most useful when it accommodates problem solvers in seamlessly carrying out their contextually shaped approaches to inquiry, without their having to divert their attention from task to tool (Mirel, 1999). If functions for methods of analysis are built into the software but users have to dig deeply into a program to figure out how to operate these often hidden or complicated features, the program falls short in usefulness. It does not match and accommodate users' models of their work.

The first usefulness challenge, therefore, is to selectively include the right types of analytical methods and build graphics and functionality for them. Designers can only know the "right" types of methods by observing in context the practices and conventions of a profession, its roles, and domain knowledge, including the data, data structures, and values common to a domain. In my study, domain mattered. Problem solvers from different subject-matter domains tackled the same type of problem but used somewhat different analytical methods. For example, I observed service technicians in telecommunications and analysts in database and intranet management commonly troubleshoot system breakdowns. At a high level of analysis, they followed the same process-of-elimination pattern of inquiry, isolating first one potential cause then another until they found the culprit. But in their detailed analysis of isolated potential causes, they looked at different relationships between factors and did not examine similarly scaled data. Nor did they look at the same type of data peripherally related to system operations or bring these other data and associated coworkers into their inquiry at the same points in the process. In sum, each group's respective domain content, data structures and values, collaborations, and professional conventions led to different methods for the same general type of problem. For their respective methods, each group expected the program to support their examinations *in a keystroke*. Usability specialists and information designers need to identify these domain-based methodological regularities and design for them.

Even if designers succeed in matching the right graphics to the right problems, other usefulness issues related to interactivity occur. As figure 9.2 reveals, displays often consist of many dynamically linked graphics, thereby enabling users perceptually and cognitively to see their problem from multiple perspectives. This visual richness presents users with a number of possible methods of analysis, and users have to be able to hone and control them for their purposes. Problem solvers want to focus, act, and interpret, without having to sort out whether the methods that they intend to use are available and what complicated steps these methods may require. Users want and expect to carry out their desired methods and see results in a keystroke.

To support this access and achieve this interactivity, interface design needs to be tightly coupled with the content of users' work. Usability specialists and information designers have to know enough about people's work in context to design for the main methods that they use in their domain, profession, and job role for a given type of problem at a likely point in their pattern of inquiry. In addition, when methods involve querying, designers also have to know and design for the values that people want to see as results in the formats in which they want to see them. Usefulness is about the substance of users' work as much as it is about structure and design.

Sample Design Situation

To exemplify the support that users need for carrying out the analytical methods of their trade, I draw on the demonstrated practices and needs of 27 retail category managers in 11 work settings. I studied these analysts doing their everyday work in their companies for at least 4 hours a session, observing them as they carried out such complex software supported inquiries as "what mix of products should I stock to maximize profits and market share?" and "what new product offering is likely to succeed in this market?" The sample design situation described next involves analysts investigating the latter question.

All of the category management analysts were specialists in analyzing complex problems. All were recognized by their managers as domain experts and were experienced in software-supported problem solving. For their analysis of product and market performance, they often worked with tens of thousands of records. All participants had at least 2 years experience interacting with data analysis and graphics software to solve complex problems, but none had experience with the new software being used in this study. The software with which they worked was first a prototype and then a beta version of interactive information visualizations designed to facilitate exploratory inquiries into complex sales, marketing, and financial questions. I observed analysts as the usability specialist and user researcher for the software products.

In seeking to discover the combination of attributes with the greatest market potential to succeed as a new product offering, category management analysts from different companies and with different products relied on the same core methods of analysis. These methods were their "tools" for structuring inquiry and imposing order and meaning on the unwieldy amounts of data that they analyzed. One method common across contexts and product categories was to identify success factors for the dominant players in a category and to use these factors as a baseline against which to evaluate one's own actual or potential product offerings as well as one's competitors'.

Central to this baseline comparison method was the need to identify the dominant players in the first place—and to identify them at all levels of the channel, product, and market hierarchies. Analysts needed to quickly see, for example, in the food, drug, and mass merchandise distribution channels, what groups of vendors, brands, products, and product attributes had a combined market share of 80% in various regions, markets, and stores. Because traversing these hierarchies and keeping track of findings and measures of success to find dominant players were complicated, analysts strived to identify dominant players as efficiently and straightforwardly as possible. They wanted in just one or two short moves to be able to eyeball the data, see at a glance the most likely dominant players, select them, see their combined market share, and continue to select until they reached the percentage that reflected dominance in a given category (this percentage varies by category).

In my study, an early prototype that category managers received did not support them in conducting this analysis-in-a-keystroke. Developers and designers did not build it in because initially they did not know enough about how this group of professionals organized their analyses or about how the multiple hierarchies and hierarchical levels of data in this field complicated these analyses. Once the prototype became part of analysts' everyday work, analysts immediately hit an impasse. They could not interactively select and get immediate, aggregated results on market share for a group of items. Their investigations literally came to a halt.

The impasse came fairly early in an inquiry. For the problem of determining a new product offering, analysts began by skipping over the vendor level data and looked at brands in selected markets to find the dominant players in the market. They displayed the sales data for all the brands in a bar chart, arranged the bars from highest to lowest sales, and selected the three brands with the highest sales. At this point, they needed to know if this selected subset dominated in combined market share. The accuracy and completeness of the rest of their inquiry hinged on picking all the brands that dominated the market. Because, with the early prototype, users could not easily get results on combined market share they were uncertain about whether this group of brands was the right group to use for the baseline. Without this certainty, they pursued no further inquiry.

Had they been able to analyze combined market share in a keystroke, they would have proceeded by funneling down in the hierarchy to these dominant brands' products and among these products identifying the best selling items, again targeting those with a combined market share that reflected market dominance. Next, for the selected products, they would have examined attributes and combinations of attributes with high sales and high market share. In the end, this funneling would reveal to them the compositions of winning products, and these compositions would serve as models for their new product offering.

Lacking the analysis-in-a-keystroke that they needed, analysts halted their inquiry and did not try workarounds because alternatives seemed too complicated and threatened the continuity of their line of reasoning and their professional conventions for structuring their inquiry. It was not just that the mental math was overloading. It was that they were unwilling to repeatedly do this mental math at every point in the product and market hierarchies. Conditions, constraints, and possible actions across the work domain and technology levels of the work space were incompatible. The functionality of the program (technology level) overly constrained users' professional practices and desired methods (work domain level) and pushed users (subjectively) beyond their cognitive limits and motivational willingness. These incompatibilities made the prototype unacceptable to users.

Once the software project team recognized how critical it was to provide this analysis-in-a-keystroke, developers built a feature for calculating the total market share of selected items, and designers created a display that gave users ready access to this information in a form meaningful to analysts in retail category management (figure 9.3).

It could be argued that the improvement took the form of a "procedural feature" not so different from support often designed for rule-driven tasks. But the way of arriving at this requirement—observing users at work and discovering domain-specific methods—and the way in which this method serves as a regularity amidst a wide array of possible paths and large volumes of data reveal that, at a deeper level, the "feature" is a structuring device for the analysis that category managers conduct. They need to know information on total market share repeatedly for items they progressively select, revise, and reselect. They are anchored by this information as they move through the product and market hierarchies. Without it they founder. It moves their inquiry steadily toward one of their core methodological patterns—setting a baseline for success and comparing the performance of other products against it.

With the new version, these users' inquiries turned around. Where they had been halted earlier, they now moved ahead easily and productively. Built-in intelligence was not needed. But design team intelligence

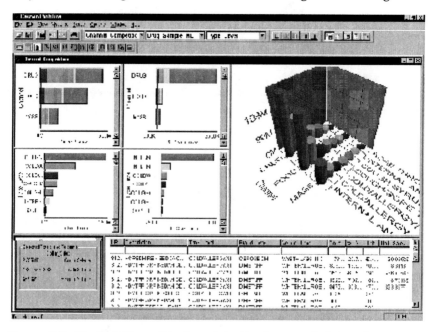

Figure 9.3. Market Share Displayed. In the bottom left hand corner, the display now has a Totals Table that provides an automatically calculated market share of all items combined that an analyst selects. The total changes whenever an analyst makes new selections.

about the content of users' work and the interactions between conditions, constraints, and actions was.

Conclusion

This discussion of designing for usefulness has revealed that complex problem solving involves adaptability, emergence, and serendipity, but it also has underlying regularities. Information designers and usability specialists need to apply design frameworks and strategies that embrace and capture these seemingly paradoxical qualities. Features for step-by-step procedures and ease-of-use strategies may be sufficient for well-structured, rule-driven, context-free tasks, but they do not provide adequate support for complex problem solving. In complex problem solving, discrete actions and procedures per se—though obviously integral to success – are not the primary design issue. It is not possible to capture a "best system" —a preset sequence of actions—for tasks that, by definition, are rarely performed in the same way twice.

The trickiest aspect of identifying and designing for problem-solving regularities is finding the appropriate level for analysis (Holland, 1998). For usefulness, regularities exist at the level at which *users* define their work-in-context and "chunk" their work processes. The best way for information designers and usability specialists to discover this user perspective is by immersing themselves in users' world of work—its contingencies, priorities, and trade-offs. It is insufficient to rely predominantly on logic, theory, second-hand descriptions from subject matter experts, or even years of experience as a designer.

To offer support for complex problem solving, software displays need to evoke a bounded freedom of interactivity shaped by context and purpose. But to build such software, designers must be guided by representations of users' work that highlight adaptability and freedom of choice. As I have argued, these representations need to stress interactions between conditions, constraints, and possible actions. When interactions between conditions, constraints, and possible actions become the focal points in representing users' work, design teams are able to see dynamic interdependencies that occur across levels of users' work space and their consequences. They can trace the ways in which the trajectory of problem-solving behavior is set by multiple conditions and constraints across different levels of the work space. When the constraints across levels compatibly interact and mutually give rise to possible actions that satisfy users' goal-oriented needs, problem solving and decision making go smoothly. However, if technological conditions impose constraints that are incompatible with those in the other levels, problem solving will falter. For example, users may need to gather data from several sources, standardize them, and relate multiple factors at once. If the infrastructure does not permit ready transmission of data from one source to another or if the program does not enable users to display and standardize data for analysis purposes, their pattern of inquiry cannot be put into play easily, perhaps not at all.

Beyond program and interface designs other approaches do exist to minimize the incompatibilities that designers may discover when they analyze interactions between conditions, constraints, and possible actions. These other interventions are outside the scope of this chapter, centering as it does on designing *software* for usefulness. Yet usefulness also can be advanced by redesigning organizational standards of practice or by restructuring work groups' communication processes and incentive systems. Such initiatives did take place in some of the projects in my study—with varying success depending on the site. The power of an interaction-oriented framework for representing users' complex work is that it shows that support for usefulness can occur at many levels of a work space. Design teams need to be acutely aware that the designs necessary for seamless and productive complex problem solving can enter the picture at any of these levels.

When designs for usefulness enter at the software level—the focus of this essay—this discussion has revealed that improvements target program scope, functionality, and architecture, not just interface design. Designing for usefulness, therefore, implies that information designers and usability specialists need to bring their immersive understanding of problem solving-in-context to bear on these upfront decisions. To do so, they must conduct contextual inquiries and create effective representations of users' demonstrated work early and make convincing recommendations for these front-end decisions.

These ideas imply major changes. Designing for usefulness often calls for the need to recast many existing development processes and the need for developers, architects, system engineers, and project managers to see users' problem solving in new ways. Representations of work that capture the dynamic, emergent, adaptive qualities of complex problem solving is a start for changing these software team members' perspectives. From representations that stress interactions between conditions, constraints, and actions, team members will be able to see clearly when incompatibilities across conditions of work are imminent in certain choices about software design and scope. Once programmers realize these incompatibilities and their negative consequences for usefulness, they should be responsive to alternatives. No programmer wants to build software that is bound to be "un-useful." Beginning software projects with an emphasis on designing for usefulness will help to avoid this result.

Acknowledgments

Many thanks to Ann Blakeslee, Russell Borland, and Dave Farkas for their very helpful comments and suggestions on drafts of this chapter.

References

ACM Transactions on Computer-Human Interaction (2000). Special issues on Directions for the New Millennium, 7(1 & 2).

Agre, P. (1997). *Computation and human experience*. Cambridge, England: Cambridge University Press.

Agre, P. (1999). The architecture of identity. *Information, Communication and Society* 2(1), 1-25.

Autros Point of Care Medication Management System, Available online: www.autros.com.

Barnard, P., & May, J. (1999). Representing cognitive activity in complex tasks. *Human-Computer Interaction, 14*, 93-158.

Bereiter, C. (1991). Implications of connectionism for thinking about rules. *Educational Researcher, 20* (April, 3), 10-16.

Bertin, J. (1999). Graphics and graphic information processing. In S. Card, J. Mackinlay, & B. Shneiderman (Eds.), *Readings in information vizualization: Using vision to think* (pp. 62-65). San Francisco: Morgan Kaufmann.

Beyer, H., & Holtzblatt, K. (1998). *Contextual design: Defining customer-centered systems*. San Francisco: Morgan Kaufman.

Burns, C. (2000). Putting it all together: Improving display integration in ecological displays. *Human Factors, 42*(2), 226-241.

Card, S., Mackinlay, J., and Shneiderman, B. (Eds.). (1999). *Readings in information visualization: Using vision to think*. San Francisco: Morgan Kaufmann.

Carroll, J. M. (2000). *Making use: Scenario-based design human-computer interactions*. Cambridge, MA: MIT Press.

Carroll, J. M. (Ed.). (2001) *HCI and the millennium*. Reading, MA: Addison-Wesley.

Cleveland, W. (1993). *Visualizing data*. Summit, NJ: Hobart Press.

Cole, M., & Engestrom, Y. (1993). A cultural-historical approach to distributed cognition. In G. Salomon (Ed.) *Distributed cognitions: Psychological and educational considerations* (pp. 1-46). Cambridge, England: Cambridge University Press.

Dillon, A., & Vaughn, M. (1997). It's the journey and the destination: Shape and the emergent property of genre in evaluating digital documents. *New Review of Multimedia and Hypermedia, 3*, 91-106.

Dillon, A. (1994). *Designing usable electronic text: An ergonomic analysis of human information usage*. London: Taylor and Francis.

Hollan, J., Hutchins, E., & Kirsh, D. (2000). Distributed cognition: Toward a new foundation for human-computer interaction research. *ACM Transactions on Computer-Human Interaction, 7* (June), 174-196.

Holland, J. (1998). *Emergence: From chaos to order.* Reading, MA: Perseus Books.

Horvitz, E. (1999). Uncertainty, action, and interaction: In pursuit of mixed-initiative computing. *Intelligent Systems, 14,* 17-20.

Human Factors (2000). Special Section on Studying Cognitive Systems in Context. *42*(1).

Hutchins, E. (1995). *Cognition in the wild.* Cambridge, MA: MIT Press.

IEEE Computer Graphics and Applications (1997). Special Issue on Information Visualization. *17*(4).

Kohn, L., Corrigan, J., & Donaldson, M. (1999). *To err is human: Building a safer health system.* Washington, DC: National Academy Press / Institute of Medicine.

Kulak, D., & Guiney, E. (2000). *Use Cases: Requirements in Context.* New York: ACM Press

Lieberman, H. (1995). Letizia: An agent that assists web browsing. *International Joint Conference on Artificial Intelligence,* Montreal. San Francisco: Morgan Kaufmann.

Maes, P., & Shneiderman, B. (1997). Direct manipulation vs. interface agents: A debate. *Interactions* IV (6).

Marchionini, G. (1995). *Information seeking in electronic environments.* Cambridge, England: Cambridge University Press.

Mirel, B. (1999). Usability test results for information visualizations: Determinant of usefulness for complex business problems. In M. A. Sasse and C. Johnson (Eds.), *Human-computer interaction INTERACT '99* (pp. 70-78). Amsterdam: IOS Press.

Morse, E., & Lewis, M. (2000). Evaluating visualizations: Using a taxonomic guide. *International Journal of Human-Computer Studies, 53,* 637-662.

Nardi, B. (1997). Activity theory and human-computer interaction. In B. Nardi (Ed.), *Context and Consciousness* (pp. 7-16). Cambridge, MA: MIT Press.

Pentland, B., & Rueter, H. (1994) Organizational routines as grammars of action. *Administrative Science Quarterly, 39,* 484-510.

Rasmussen, J., Pejtersen, A., & Goodstein, L. P. (1994). *Cognitive systems engineering.* New York: Wiley.

Root cause analyst, Available online: rootcauseanalyst.com / software.htm.

Senay, H., & Ignatius, E. (1990). Rules and principles of scientific data visualization *(Technical Report GWU-IIST-90-13),* Institute for Information Science and Technology, Available online: www.siggraph.org / education / materials / HyperVis / percept / visrules.htm.

Shneiderman, B. (1996). The eyes have it: A task by data type taxonomy for information visualizations. *Proceedings of IEEE Workshop on Visual Languages '96*, 336-343.

Spiro, R., & Jehng, J. (1993). Cognitive flexibility and hypertext theory and technology for nonlinear and multidimensional traversal of complex subject matter. In D. Nix & R. Spiro (Eds.), *Cognition, education, and multimedia* (pp. 163-205). Hillsdale, NJ: Lawrence Erlbaum Associates.

Suchman, L. (1997). Centers of coordination: A case and some themes. In L. Resnick, R. Saljo, C. Pontecorvo, & B. Burge (Eds.), *Discourse tools, and reasoning: Essays on situated cognition* (pp. 41-62). Berlin: Springer-Verlag.

Sutcliffe, A., Ryan, M., Doubleday, A., & Springett, M. (2000). Model mismatch analysis: Toward a deeper explanation of users' usability problems. *Behaviour and Information Technology, 19*(1), 43-44.

Tukey, J. (1977) *Exploratory data analysis.* Reading, MA: Addison-Wesley.

Tweedie, L. (1997). Characterizing interactive externalizations. *Proceedings of CHI '97, ACM Conference on Human Factors in Computing Systems*, Atlanta, 375-382.

Vincente, K. (1999). *Cognitive work analysis.* Mahwah, NJ: Lawrence Erlbaum Associates.

Vincente, K. (2000). HCI in the global knowledge-based economy: Designing to support worker adaptation. *ACM Transactions on Computer-Human Interaction 7*, 263-280.

Vincente, K., & Rasmussen, J. (1992). Ecological interface design: Theoretical foundations. *IEEE Transactions on systems, Man and Cybernetics 22*, 589-606.

Ware, C. (2000). *Information visualization: Perception for design.* San Francisco: Morgan Kaufman.

Wright, P., Fields, R., & Harrison, M. (2000). Analyzing human-computer interaction as distributed cognition: A resources model. *Human-Computer Interaction, 15*, 1-41.

Ziech, K. (2000, April). Personal conversation.

10

Complex Problem Solving and Content Analysis

Michael J. Albers
University of Memphis

There is much written about the vast quantity of information available and the problems of information overload. The amount of information available on any normal size corporate web site, especially the intranets, has long surpassed the amount a person can digest and analyze.

This problem becomes even more complex as users expect to take the information and use it to accomplish something. Because users rarely need information based on simple look ups (field X has a value of 5, so the answer is no.), we almost always must help users solve a complex problem. Unfortunately, our design and presentation methods are struggling to evolve into methods to effectively help sort and define the requirements for understanding the preponderance of information in specific contexts. A straightforward process to provide users with complex information in a useful format does not exist. In solving the complex problem, the potential choices and reasons for making the choice become the dominate factor. Only after we understand these particular choices and reasons can a problem-solving strategy be developed. That problem-solving strategy then becomes the basis for the information design (Rasmussen, 1986).

In complex problem solving, rather than simply completing a task, the user needs to be aware of the entire situational context in order to make good decisions. In web-based knowledge management systems and other information systems designed to support complex problem solving, the early analysis must uncover the user goals and information needs in order to allow the user to build the required picture of the entire situation. The information design goal is very black and white. The information must frame the problem as a business issue, provide access to all relevant information and eliminate all irrelevant information. Nothing else will work, nor should it. Methods of achieving that goal are, even understated, much murkier.

As used within this chapter, complex problem solving is the problem solving and decision making used in ill-structured situations to gain awareness of the problem, analyze a situation, and make decisions, and track the results. The converse, simple problem solving, applies to situations which have defined solution paths. I should point out that this definition is neither as precise as used in cognitive psychology nor quite the same as used by designers of decision support software.

Most of the work on web design has focused on information architectural or navigational issues, which are important, but they often ignore the issues of web-based content itself (McGovern, 2001). Information architecture's focus has been defining the layout and categorizing information in systems. Granted, making sense of information always requires a certain level of layout and categorizing, but the text content must drive the design problem. Providing information content in ways that makes sense to users and allows gaining an understanding of the situation is always important, but is absolutely necessary for systems that support complex problem solving. For the information to be useable, the content in the context of the user's situation must drive both the information architecture and the information presentations.

Providing the knowledge appropriate to the situation means designing from a detailed understanding of the user environment and user goals. That understanding can then be applied to the information design.

> This broader view is necessary to capture the following traits of complex tasks: paths of action that are unpredictable, paths that are never completely visible from any one vantage point, and nuance judgments and interpretations that involve multiple factors and that yield many solutions, each with costs and benefits. (Mirel, 1998)

The basis for understanding which information to collect and how to organize it to properly portray the situation can be considered as analogous to the three levels of situation awareness, which operates on three levels (Randel & Pugh, 1996):

- **Knowing the data.**
  What elements make up the important data elements and what are their current values.

- **Knowing the interrelations within the data (information).**
  How do the data elements interact? What data elements should be viewed as a coherent whole to ensure building the proper picture.
- **Being able to make predictions or relate the information to the larger picture (knowledge).**
  How are the data elements expected to change in the future based on their interrelationships and on any interventions caused by the user.

Content analysis, presented in this chapter, provides a method of ensuring that the information within a system allows the user to achieve the foregoing criteria. It also ensures that the designer has enough situational knowledge to present the information in a manner that fits Marchionini's (1995) three dimensions of information: specific to the situation, in the proper quantity, and presented in a timely manner. As a result, readers can easily acquire, understand, and use the required information to enhance their understanding of the situation and to make decisions. Consider the contrast of this view of easily acquiring and relating information with the normal search engine approach of dumping a list of possible sources for the user to sort out.

Clarification of terms

Understanding the viewpoint of this article requires clarification of few points.

I am discussing user goals and information needs in the context of ill-structured but defined situations. The report analysis performed on a monthly basis by a financial analyst or a customer making a buying decision for a major purchase fits within this category, but solving novel or one-time problems does not. Similar concepts have appeared in the literature: Vincente (1999) uses the term *situated context*, Suchman (1987) uses it *situated action* and Roth, Bennett, and Woods (1987) use *unanticipated variability*.

1. I am not discussing expert system design or other systems that by design automatically provide an answer to the user. The root problem is that expert systems privilege abstract, rule-based knowledge and attempt to solve the problem internally, as opposed to providing knowledge to the users (Raybould, 1998; Winsor, 1996). Abstract, rule-based knowledge has proven less than useful with ill-structured problems. Also, moving away from a simple question–answer dialog requires natural language input, which still has significant problems (Smeaton, 1994). This chapter considers how computers support people in

problem solving; it does not consider how computers can replace them or directly provide the answer. Computers and people both excel at different tasks, effective design must balance the two and let each work at the tasks which they do best.

2. Although drawing upon the situation awareness research, the problem-solving focus lacks the immediacy of response that situation awareness stresses, but rather a longer term, more strategic effort where the user has time available for making, implementing, and tracking the decision. Situational context, as described in this chapter, works at a longer term, more strategic level, rather than the second-by-second focus of the situation awareness literature.

3. Only recently, with technologies such as XML and dynamic web page generation, has interaction design gained the underlying tools to actually support complex problem solving in the manner I call for here. With XML, the reports can be marked-up and the various elements presented to the user grouped together in a manner which supports the information needs. Unfortunately, how to use those technologies to help us get from where we are to where we need to be has not been adequately addressed. Instead, technology issues have dominated with human issues relegated to a second tier. Most of the current research seems to be focused on computer-to-computer transactions or simply on the techniques of getting XML data into and out of a database rather than on methods of context generation. As an immature technology, research into all aspects of XML are needed, but we must not forget that the real usefulness lies not in pulling data out of a database, but in using it afterwards in a manner which fits the user's real-world goals and information needs.

Examples of simple and complex problems

The bank ATM is a common example in many books on task analysis and HCI. It provides a good example of simple problem solving. It has a small, fixed set of tasks and each can be fully defined. The ability to fully define each task and provide one path qualifies it as simply problem solving.

On the other hand, determining problems with corporate sales is complex problem solving. The user may use a set of guidelines (heuristics) for resolving a low sales problem, but these must be adjusted for each individual problem. Each new piece of data the user uncovers affects the path taken and the eventual outcome. The sales analysis qualifies as complex because it does not lend itself to being performed with a defined set of tasks nor can those tasks be performed in a fixed order.

For a sales example, let's assume the monthly sales figures show a drop of 10% from the previous month. Viewed out of context, a drop of this size seems to call for immediate action. However, perhaps these are the figures for January retail sales and the normal decrease from December is 14%. Or the sales are for a northern store that was closed over a weekend because of a major four-day snowstorm. Or the advertising flyers mailed late because of printer problem. None of this information will appear on the sales report itself, yet making a reasonable decision about the numbers requires the user to know about them.

Structured versus ill-structured problems

The open-ended nature of complex problem solving complicates defining the information requirements. In contrast to the step-by-step analytical process of task performance, problem solving "rarely arises straightforwardly, but rather results from a long and recursive process with backtracking and erratic switching among the following activities: thinking about ideas, production, reorganization, modification, and evaluation" (Nanard & Nanard, 1991, p. 50). Supporting systems for complex problem solving requires building on the way in which people rapidly assess situations and make decisions based on theories such as Klein's recognition-primed model rather than for classical decision models or task analysis such as GOMS (Albers,1996). Klein's (1999) recognition-primed decision making model brings out the rapid evaluation and assessment skilled problem solvers use. Real-world decisions get made based on choosing the first relevant solution and "muddling through" (Klein, 1999; Hollnagel, 1993). Support for complex problem solving must focus on the user's goals and needs and provide that information in a manner that allows a person to rapidly develop a solution appropriate to the situation (Albers, 1997; Mirel, 1998, this volume).

In a highly structured environment, the user's basic goal is essentially one of efficiently completing the task. A step-by-step route can be predefined as the correct path to an answer and that path can be supported and enforced by a computer system. In a well-defined domain, the user approaches are limited and can be fully defined by the designer.

On the other hand, in the ill-structured environment, the user's goal is one of analysis and problem solving. The step-by-step route to completing a task simply does not exist. In an ill-structured domain, instead of following a set path, the user continuously adjusts their mental path as new information presents itself. As a result, each user takes a slightly different path and the designer can't assume that an understanding of how one person performs the task describes anyone else. The designer can't even assume that the information needs are consistent between users or which information a user will view before making a decision. Yet, the designer is tasked with creating a design that provides the information when and how the user wants it.

To support the problem-solving needs of an ill-structured environment, the system must present the new and old information and assist in integrating that information into the user's mental model to support reaching a correct decision. As such, the system must do more than just contain the information; it must assist in problem recognition and provide the information to define the situation and support the decision. It must then support problem follow-up to verify a correct decision was made and the overall situation is proceeding in the expected manner.

Support for complex problem-solving

Being able to look up information and then compare or compile information must be a fundamental design aspect when addressing complex problem solving. However, most information design does not focus on providing answer to complex problems. Even user tests often reveal a bias by focusing on answering questions with simple answers, "What was total sales in the eastern region in March?" A bias that definitely makes the test data easier to quantify. Yet, a sales manager rarely actually needs this single piece of information. Instead, the sales manager wants to understand what factors contributed to the eastern region sales in March, how they compared to previous months, and what can be done to increase those sales. The sales manager is engaged in complex problem solving.

Business report analysis is one area of complex problem solving that can highly benefit from the design ideas put forth in this chapter. Consider the analysis of a corporations financials prior to making an investment. Currently, the financial analyst would receive a copy of the annual report as well as other related supporting documents. Then, bits and pieces need to be pulled from each report, calculations performed, and compared against numbers that are in other pieces of the reports. In a similar vein, consider a typical user trying to make investment decisions. A wealth of information about a company exists, but finding and putting it together into a coherent image requires skills the average investor lacks.

To create an effective information design, the designer must understand what the user's real-world goals and information needs are. For that, we can use context analysis. Notice that the goal of providing quality content does not mean to automatically extract and calculate various financial ratios. Although some of these ratios can be calculated automatically, the real goal of the business analyst is to understand the corporation financially, not view financial ratios. Their training and experience are not focused on calculating ratios, but on comparing–contrasting the results and making decisions based on those ratios. Early design analysis must define the user's common goals, the questions that must be answered to achieve those goals, and the information that answers the questions. It must go beyond the simple result and determine what problems or confounding data might be buried in the numbers.

Of course, the actual design implementation requires extensive prototyping work and careful consideration of cognitive loading to prevent information overload. All the information possibly relevant to support addressing the user's goals and information needs simply cannot be dumped onto the screen. Nor can a hierarchy of links solve the design problem of effective information presentation.

Orasanu and Connolly (1993) discussed how, while research often looks at decision making in isolation, in reality, decision making occurs as part of larger tasks and makes up only a single element in achieving a larger goal. They place decisions within a cycle which "consist[s] of defining what the problem is, understanding what a reasonable solution would look like, taking action to reach that goal, and evaluating the effects of that action" (p. 6). Terveen, Selfridge, and Long (1995) looked at the problem of helping to manage the knowledge needed to make decisions and revealed the interrelated knowledge issues to consider.

> The pragmatics of knowledge use are critical. Simply recording a factor is not enough; issues such as where in the process knowledge is to be accessed, how to access relevant knowledge from a large information space, and how to allow for change also must be addressed. (p. 3)

Defining users' information needs in a manner that captures the pragmatics of the knowledge presents a major problem. Too many systems are designed with an assumption of providing single data points or addressing each issue independently, or with the idea that as long as the user has the data and data manipulation tools, the system is acceptable.

Issues with information design and complex problems

Clearly addressing the information design issues associated with helping users solve complex problems requires focusing on user goals and information needs. This section of the chapter explains how the following three points relate to addressing complex problems.

- Understanding the user goals and information needs
- Understanding information in context
- Understanding the situational context

Understanding user goals and information needs

In a complex problem-solving environment, attempts to describe step-by-step actions break down because no single route to a solution exists. No matter how detailed the analysis, a step-by-step description for solving sales report problems cannot be obtained. A conventional task analysis would reveal the basic steps and calculations used, but doesn't do a good job of capturing the underlying reasons that drive performing the actions or the information relationships used to analyze problems found within a report. The common problem with task analysis is that it captures what the user does but fails to capture what motivated the user to perform the action.

In complex problem solving, people don't follow the nonexistent fixed path, instead they continually adjust their mental paths as new information presents itself. The nonfixed path arises from the nonlinear aspects of problem solving. Addressing the mental shift to nonlinear thinking breaks with the fundamental philosophy of step-by-step analysis. To support complex problem solving, the interface must place the information within the situation context and allow user to develop and maintain an awareness of the situational context as it changes over time. The content analysis reveals the general user goals and information needs and the potential subgoals that must be addressed as they proceed to a resolution. The problem is not impossible, it simply requires a different viewpoint and mindset on the part of the system analysts and information designers.

Instead of a linear step-by-step procedural process, providing information for complex problem solving could be better viewed as an process comparable to sintering ceramics. In sintering, a fine power is compressed and heated. Without actually melting, the individual grains join together to form a solid. In complex problem solving, the individual pieces of data need to be compressed and manipulated to form a coherent picture of the real-world process they describe. An in-depth study of both sintering and complex problem solving requires studying the situation at all levels. At the microscopic level, to see how the individual elements group together and adhere. At the larger levels to see how stress affects the result and how separate tasks merge to form a complete solution. Although the user may not have to mentally consider each of the levels, the information designer must address each during the early analysis and understand them.

Understanding information in context

A decision reached in a problem-solving situation is based on how people interpret the information. People bring to the system a real-world goal of obtaining information to help solve a problem, mentally forming the relationships within the information, relating it to their real-world situations, and, most importantly, using it to perform a useful/correct action (Belkin, 1980). The design must fit the information and convey the knowledge appropriately (Laplante & Flaxman, 1995; Tufte, 1983).

All too often information is not designed for integration with other information, but rather is optimized for its own presentation. Many hypertext systems assist in presenting complex information, but Johnson-Eilola and Selber (1996) argue that, rather than exploiting the computer environment by using hypertext, most hypertexts tend to maintain the traditional hierarchical organization of paper documents. As a result, the user suffers from a high cognitive workload caused by inefficient information presentation. Users have a hard time remembering or considering subtle cause and effect, goal-oriented relationships that exist between the information being viewed and other relevant information. Web browsers contribute to the lack of access to integrated information because they operate as simple viewers into data but provide no support for integrating or analyzing the data to obtain information from it. An effective content analysis should provide for understanding the relationships which build a solid foundation for giving advice to solve a problem (Casaday, 1991) and to placing the entire situation in context (Endsley, 1995).

In the end, if the user cannot find or develop the information relationships, the online information fails to fulfill its expectations because it fails to anticipate the users' real-world needs. "Decision makers address complex problems for which no ready-made answers exist in any database. Thus it cannot be expected that the relevant information be found by direct means, but inferred" (Waern, 1989, p. 172). If the user cannot find or develop the information relationships (achieving the second step in understanding information), the web-based information fails to fulfill its goals and does not meet the users' real-world needs.

Understanding the situational context

Two major factors complicate understanding the situation: knowledge limitations about the situation's current state and response, and cognitive limitations for mentally handling the available information integration (Wickens, 1992; Yoon & Hammer, 1988). A problem with many existing (and failed) systems was that the designers applied a technological solution which provided data, but never defined the information needs and relationships required to understand the situation. Post hoc error analysis reveals that most interactive system errors can be traced to the fact that the data does not accurately represent real-world conditions. The system provided complete and accurate data, but did not provide adequate context to allow the user to create a comprehensive picture of the situational context (Macaulay, Fowler, Kirby, & Hutt, 1990).

People identify a problem and make a problem-solving decision based on how they interpret the information around them. Unfortunately, the interpretation often has less to do with reality and more to do with what the person expects to see. Strong expectancy biases in data interpretation must be expected (Klein, 1988). Also, the presentation format exerts a strong influence on how people interpret the data. Some research has found different presentations can actually reverse people's decisions while using the same data (Johnson, Payne, & Bettman, 1988; Tversky & Kahneman, 1981).

Building on the way in which people rapidly assess situations and make decisions requires considering information requirements in the light of models such as Klein's recognition-primed model and considering how the context drives the decision-making process. Besides the information itself, the problem context contains many interrelationships unique to the situation that distinguish it from other related contexts. The key to understanding the situational context requires understanding these relationships. The interrelationships within the information must be considered as part of the design, because only by understanding these relationships can the user obtain an acceptable view of the situation.

Why content analysis is needed

Earlier, I claimed that understanding the situation required the system to allow the user to accomplish three different levels of understanding: knowing the data; knowing the interrelations within the data; and being able to make predictions or relate the information to the larger picture. This section expands upon these three concepts and explains their relevance to content analysis. The next section examines how a designer can ensure the content analysis captures this information.

Knowing the data

A problem with many reporting tools lies with a design that treats all information as equal and delegates to the user the responsibility of integrating the data into a coherent collection of information. The recipients of the report are expected to sort out the data by picking and choosing what is relevant, combine it with data on other reports, and make sense of it themselves. The context of information use never entered into the design.

This method of dumping data upon people and hoping they can sort it out (the basic idea behind current Internet search engines) fails because it does not contribute to letting the person get a grip on the information so they can turn it into knowledge or to providing effective information presentation.. Likewise, dumping report information online fails to account for the dynamic, interactive potential of the on-line environment (Heba, 1997). Thankfully, most designers have moved beyond the stage of calling it sufficient to simply present data to the user.

The information design must ensure that information salience is proportional to its importance. All data is not created equal. Some data is more relevant to the situation than other data. The design must not allow easy to present but unimportant data to dominate.

Knowing the interrelations within the data

Effective information design for these situations requires focusing on the interrelationships between information and how those interrelationships fit into the user's mental model of the situation. The issue of interrelations cannot be swept aside with arguments that a skilled decision maker knows about the relationships and will look for them. True, the skilled decision maker knows about the relationships, but too many factors can contribute to them not being seen. Factors such as time pressure, desire for closure, expectancy bias caused by external pressure, simply forgetting, or having a design that makes grasping the relationship too complex. When solving a problem, the user's initial difficulty is recognizing what is wrong and what information is needed to verify the problem and suggest possible solutions. The user may not either know what available information relates to the problem or, because of cognitive tunnel vision, may not think to look at pertinent information.

The design difficulty arises because in solving the complex problem, a person must evaluate and integrate information that occurs at multiple locations (some of which may not be web based). The vital pieces of information are often dispersed in single sentences or paragraphs located on multiple web pages. Cognitive constraints prohibit a person from mentally tracking this information and current web browser technology makes it hard to compare–contrast the information. As a result, confusion often arises over different possibilities and redundant analysis occurs resulting in substandard decisions.

Multiple studies on how people solve problems in the real world reveal a consistent feature of people constantly adjusting their goals and sub goals to allow for the dynamic nature and quality of the information available (Klein, 1999; Orasanu & Connolly, 1993). Unfortunately, people have a hard time integrating information and relating various data points to each other. Inasmuch as people have difficulty integrating information and relating various data points, they have a hard time effectively using the information (Wickens, 1992). Unless it is properly organized, humans cannot efficiently process large amounts of information, especially with the incomplete information available in a real-world context (Gerlach & Kuo, 1991; Woods & Roth, 1988). It must be arranged and presented in a manner supporting the users' goals and information needs. Also, they have a hard time remembering or considering subtle cause and effect relations that exist between the information being viewed and other information. Cognitive tunnel vision can cause them to ignore and forget to consider other information. Also, different presentations cause different approaches to the solution (Elam & Mead, 1990).

As such, the information designer walks a thin line of figuring out to present information and the relationships while not overloading the user.

Being able to make predictions or relate the information to the larger picture

After understanding the relationships within the situation, the person needs to use the data and relationships to do something. Essentially, they must take the information provided and by predicting the future effects of various decisions, choose one. The content analysis must address potential solutions and how they will effect the situation.

A primary factor complicating predicting the effects of changes arises because complex situations are highly dynamic and usually exhibit a nonlinear response to changes. But people make linear extrapolations and are extremely poor at judging nonlinear data. Thus, the system must do more than just present information and must even do more than just assist in making the decision, but must continue past the decision point and assist in tracking all the changes that result from the decision and prevent cognitive tunnel vision. Many failures occur because of the assumption that the initial conditions would somehow suddenly jump from initial to final state. Unfortunately, people find it difficult to extrapolate information in nonlinear systems; they tend to assume linear response or underestimate the amount of change. In the short term, the extrapolation can be within acceptable limits, but over time the differences become huge. A design goal must be to prevent the problem situation from moving off without the user knowing about it.

Unfortunately, a second major factor impacting predictions arises because a real-world system has many external factors that will not be captured in the system. These make the system unpredictable and greatly complicate the analysis. They also provide the incentive to engage in the content analysis of this chapter. With their susceptibility to external effects, they no longer have a simple one task–one path relationship. The external influences can change the proper decisions and information requirements, thus the analysis must consider what changes might occur and when and why.

What content analysis must accomplish

Information always exists in the overall information structure within the organizational context. The major distinguishing factors that drive the effectiveness of the information depends on how well the information addresses the following four questions.

- How does the person expect to access the information?
- What do they need to gain a clear picture of the situation?
- What relationships exist within the information that clarify the situation?
- What information and information interrelationships provide support for tracking results?

Context analysis defines the problem domain in terms of user goals and the information required to achieve those goals. It seeks to build a model of the factors required to maximize user understanding and performance (Hix & Hartson, 1993), and addresses how information is integrated or combined in the process of making the decision. Error identification, error correction, and out-of-the-norm situations are emphasized because they require the most problem-solving support. When solving a problem, people must recognize what is wrong, what information is needed to verify the problem, and possible solutions. Thus, the content analysis must identify how people recognize the problems in a situation, how they verify it, and what information they require to propose and track a solution.

The content analysis includes both static and dynamic information. Examples of static information are procedures, policies, or rules (laws) affecting the problem, information which, for normal purposes, can be considered fixed. Dynamic information is sensitive to the situational context and varies on a short time scale. (e.g., production/sales information or employee moral).

The following considers some of the major aspects of context analysis.

- The analysis is determined by the user's goals or objectives and the information required to achieve them, not conventional tasks (Hackos & Redish, 1998; Carroll,1995).
- The goal is to determine what the user would ideally like to know, but acknowledges that incomplete information is the norm and complete information can never be provided (Blandford & Young, 1996).
- The analysis helps users conceptualize and visualize the information relationships (Treu, 1990).

- The analysis attempts to accurately represent real-world conditions (Macaulay et al., 1990).
- The analysis does not define a single "best" answer or path to the solution, but attempts to uncover all the common paths and provide the information needed to support them (Hallgren, 1997).
- The analysis does not assign fixed priority to goals or sub-goals; short-term and long-term goal hierarchies should be distinct. It is assumed that during the decision-making process, the relative priority of goals change as the users' understanding of the situation changes (Endsley, 1995).

Content analysis that supports complex problem-solving must provide an effective means of extracting the potential, although unpredictable, paths and information needs and goals of the user. As Benyon (1997) states, "we must shift attention from humans, computers and tasks to communication, control and the distribution of domain knowledge" (p. 46). From the start, the design must allow users to continuously adapt information structure while they search for solutions. Thus, the process of supporting complex problems solving is to help the users:

- Identify the important elements of the situation
- identify the relationships between the elements
- Identify the information required to track results to ensure the decision is causing the desired response.

Identify important elements

The readers bring to the system a set of real-world goals which the system design must consider from the earliest stages (Belkin, 1980). The first step in meeting people's information needs requires initially defining their goals and information needs. The collected goals and information needs create a vision of the users focused on what real-world questions they want answered and why (Mirel, 1992).

Defining the goals and information needs becomes complicated because there is no "right way" to solve a complex problem. As such, the initial analysis cannot define a single path. With complex problem solving there is no single path and no way of predicting when the user will make a decision. Each person works at solving the problem in their own way and requires information presented to fit their current needs. Also, experienced users come up with many inventive ways to use a program. The closer the final design comes to handling the dynamic aspects of the real-world situation, the more useful and accepted the final result will be. With the dynamic nature of complex problem solving and variability of the situation, slightly different starting points can require different information and resulting decisions. Thus, the requirements analysis cannot focus on defining a single path; nor can it define exactly which information is desired. Instead, the analysis must uncover potential information requirements and presentation methods (Blandford & Young, 1996).

Wickens (1992) pointed out that the major problem in decision making is understanding the problem. A decision is based on how people interpret the information around them. Incorrect or incomplete information can lead to incomplete or invalid decisions. Even with complete information, different presentations cause different approaches to the solution (Elam & Mead, 1990). Having the data available does not equal understanding. Thus, the users' main need is information that supports gaining a clear understanding of the problem and the possible solutions. During the requirements analysis, the information designer must gain an understanding of what information is relevant, how the information is obtained, how it relates to other information, and how to present the information to maximize revealing those relationships.

To accomplish this, both Endsley (1995) and Marchionini (1995) claim the analysis must identify the major goals and the subgoals necessary to meet the goals. As part of defining the subgoals, each of the major decisions needs to be identified along with the information needed to support achieving each goal.

Identify relationships between important elements

Truly understanding the situation requires understanding all the factors involved. Many situations may appear similar but have radically different causes and solutions. In analysis for complex problem solving, the user's goal and the individual data elements about the situation never exist in isolation. Understanding of the entire situation arises not from just knowing about the individual elements, but by understanding how they are currently related and how any changes will ripple through the system.

Most situations applicable to complex problem solving are highly sensitive to change. Changes in one data element can rapidly, and in a highly nonlinear manner, affect other data elements. Thus, the knowledge about the relationships between pieces of information must be captured so it can be provided to the user. Besides just defining the information needs (too often the endpoint of conventional task analysis), how that information is integrated or combined in the process of making the decision needs to be addressed.

As people address a complex problem, a structure emerges which is based on the context, the interrelationships between the important elements, and the user's schema. Highly effective support for complex problem solving thus must match the emergent structure, a structure that undergoes multiple shifts in its point of view as the problem solving progresses. Grant (1994) claimed effective design models must provide both an understanding of the context and an understanding of what causes people to switch contextual models.

Identify information to track results

As a result of executing the decision, the overall situation changes. However, there is no guarantee the changes will correspond with the anticipated results. An analysis for an information system that supports the evaluation of complex problem solving must provide more than the initial situation condition. Because of the nonlinearity issues, the initial decision probably will not be completely and successfully completed. Most real-world systems contain unintended feedback loops that must be exposed as the situation progresses. It makes no sense to expect to make a decision and allow the situation to progress until it should have reached a completion point. Thus, changes must be monitored and new decisions may be required. The user must be able to define and monitor goal points to ensure the situation is progressing in an orderly (or at least the expected) manner toward the final goal.

After making a decision, the feedback provides the basic information that lets the user know if the entire system is currently on the expected path to achieve the desired result. Feedback must be automatic and rapid and designed to provide the salient information in order to short circuit cognitive inertia and keep people following and examining the real data, rather than rationalizing all results as fitting with the desired outcomes. In systems with no or slow feedback, users are unable to make mid-course corrections because they don't know the current status. Of course, the time scale for "rapid feedback" is situation specific and can range from seconds to days or weeks.

Nonlinearity means minor variation in the initial conditions can have major ramifications later on. The impact of seemingly inconsequential nuances of data can have serious effects, as nonlinear systems are very hard to predict. For a business system that is constantly responding to multiple influences, the difficulties of making accurate predictions is obvious. The multitude of factors that simultaneously influence the process lead to the multithreaded aspects. The user must be concerned with multiple threads of information, each basically independent, but weaving around and influencing each other at various times.

The feedback also helps handle the changes occurring in the situation during the transition period between making the decision and achieving the final result. No complex system jumps from a problem state to a final problem-free state instantly. Even after the content analysis, many processes relevant to the situation will be only partially understood, specially those involving transformations and situation responses during transition between states. During the transition, many data elements are in flux, the user must not be mislead into thinking these changing elements are new problems that must be addressed.

Future research requirements

Most of the present information presentation technology still closely mirrors its paper-based roots. Only recently, with technologies such as XML, has information design gained the underlying tools to actually support complex problem solving in a general manner. With these new technologies, we finally have the means to break up the old contextual elements, recombine them at will, and present information uniquely focused on a problem. However, although the need to support complex problem solving is undisputable, we lack a clear methodology to support capturing the information requirements, reordering and dynamically constructing information to fit a situational context, and designing the interface for such a system.

Further research must provide a foundation to provide an understanding of how web design affects information analysis, and to identify the factors to consider to maximize speed and efficiency of using web sites for complex problem solving and developing clear situation context. The research needs to address how to:

- Perform the content analysis which uncovers the goals and information needs to understand the problem situation.
- Perform user testing which addresses solving a complex problem, rather than providing simple information look up.
- Identify design criteria which support how people use the web to solve complex problems.
- Define design characteristics and guidelines which aid and hinder information evaluation and developing situation context.
- Define how to incorporate the various technologies such as XML, adaptive interfaces, and data mining, into designs that provide effective information.

Conclusion

The large-scale coherence of the informational system comes from constructing the relationships between all the major pieces and lets the users synthesize the work as a whole (Thuring et al., 1995). In a properly designed system, users can:

- Rapidly access highly integrated information.
- Work in an environment which allows for dynamic modification of data format.
- Receive support for examining related information, or cause and effect relationships and to identify discrepancies between observed and anticipated behavior.
- Receive feedback to track and assess the quality of the decisions.

Admittedly, this chapter has not described a methodology for performing context analysis. Actually, the existing user-centered methodologies should be more than adequate. The importance of proper user-centered design methods, such as contextual design (Beyer and Holtzblatt, 1999), personas (Cooper, 1999), and scenario-based design (Carroll, 1995) and the involvement of users during development cannot be over-emphasized. It is not so much a need for a new set of methods as the need to have a different viewpoint while performing the existing ones that can make content analysis a viable information design technique.

Content analysis focuses on defining the content the user needs and how the user needs it presented. Only after the content requirements are defined can navigational and presentation decisions be made. Anything less than this focus short-changes the entire design. Good navigation with bad content is useless. With bad content, the user's goals and information needs cannot be met.

References

Albers, M. (1996). Decision making: A missing facet of effective documentation. *Proceedings of the 14th Annual International Conference on Computer Documentation*. Washington, DC: ACM.

Albers, M. (1997). Information engineering: Creating an integrated interface. In M. Smith, G. Salvendy, & R. Koubek. (Eds.). *Proceedings of the 7th International Conference on Human-Computer Interaction* (pp. 213-216). New York: Elsevier.

Belkin, N. (1980). Anomalous states of knowledge as a basic for information retrieval. *The Canadian Journal of Information Science, 5*, 133-143.

Benyon, D. (1997). Communication and shared. knowledge in human-computer systems. In M. Smith, G. Salvendy, & R. Koubek. (Eds.), *Proceedings of the 7th International Conference on Human-Computer Interaction* (pp. 43-46). New York: Elsevier.

Beyer, H., & Holtzblatt, K. (1998). *Contextual design: Defining customer-centered systems*. San Francisco: Morgan-Kaufmann.

Blandford, A., & Young, R. (1996, November). Specifying user knowledge for the design of interactive systems. *Software Engineering Journal*, 323-333.

Carroll, J. (1995). Introduction: The scenario perspective on system development. In J. Carroll (Ed.), *Scenario-based design: Envisioning work and technology in system development* (pp. 1-18). New York: Wiley.

Casaday, G. (1991). Balance. In J. Karat (Ed.). *Taking software design seriously: Practical techniques for human-computer interaction design* (pp. 45-62). San Diego: Academic Press.

Cooper, A. (1999). *The inmates are running the asylum*. New York: SAMS.

Elam, J., & Mead, M. (1990). Can software influence creativity? *Information Systems Research, 1*(1), 1-10.

Endsley, M. (1995). Toward a theory of situation awareness in dynamic systems. *Human Factors, 37*(1), 32-64.

Gerlach, J., & Kuo, F. (1991). Understanding human-computer interaction for information system design. *MIS Quarterly, 15*(4), 527-550.

Grant, S. (1994). Modeling complex cognition: Contextual modularity and transitions. In *Proceedings of the 4th International Conference on User Modeling*, pp. 157-162, Hyannis, MA. The MITRE Corporation.

Hackos, J., & Redish, J. (1998). *User and task analysis for interface design*. New York: Wiley.

Hallgren, C. (1997). Using a problem focus to quickly aid users in trouble. *Proceedings of the 1997 STC Annual Conference*. Washington, DC: STC.

Heba, G. (1997). Digital architectures: A rhetoric of electronic document structures. *IEEE Transactions on Professional Communication, 40*(4), 275-283.

Hix, D., & Hartson, R.. (1993). *Developing user interfaces: Ensuring usability through product and process*. New York: Wiley.

Hollnagel, E. (1993). Decision support and task nets. In G. Klein, J. Orasanu, R. Calderwood, & C. Zsambok, (Eds.). *Decision making in action: Models and methods* (pp. 31-36). Norwood, NJ: Ablex.

Johnson, E., Payne, J., & Bettman, J. (1988). Information displays and preference reversals. *Organizational Behavior and Human Decision Processes, 42*, 1-21.

Johnson-Eilola, J., & Selber, S. (1996). After automation: hypertext and corporate structures. In Patricia Sullivan, and Jennie Dautermann, (Eds.). *Electronic literacies in the workplace: Technologies of writing*. Urbana, IL: NCTE.

Klein, G. (1988). Do decision biases explain too much. *Human Factors Society Bulletin, 32*(5), 1-3.

Klein, G. (1999). *Sources of power: How people make decisions*. Cambridge, MA: MIT.

Laplante, P., & Flaxman, H. (1995). The convergence of technology and creativity in the corporate environment. *IEEE Transactions on Professional Communication, 38*(1), 20-23.

Macaulay, L., Fowler, C., Kirby, M., & Hutt, A. (1990). USTM: A new approach to requirements specification. *Interacting with Computers, 2*(1), 92-118.

Marchionini, G. (1995). *Information seeking in electronic environments*. New York: Cambridge University Press.

McGovern, G. (2001) Content is not a technology issue. *New Thinking,* June 18.

Mirel, B. (1992). Analyzing audiences for software manuals: A survey of instructional needs for 'real world tasks.' *Technical Communication Quarterly, 1*(1), 15-35.

Mirel, B. (1998). Applied constructivism for user documentation. *Journal of Business and Technical Communication, 12*(1), 7-49.

Nanard, J., & Nanard, M. (1991). Using structured types to incorporate knowledge in hypertext. *Proceedings of Hypertext '91* (pp. 329-343). Washington, DC: ACM,.

Orasanu, J., & Connolly, T. (1993). The reinvention of decision making. In G. Klein, J. Orasanu, R. Calderwood, & C. Zsambok, (Eds.). *Decision making in action: Models and methods*. Norwood, NJ: Ablex.

Randel, J., & Pugh, L. (1996). Differences in expert and novice situation awareness in naturalistic decision making. *International Journal of Human-Computer Studies, 45*, 579-597.

Rasmussen, J. (1986). *Information processing and human-machine interaction: An approach to cognitive engineering*. New York: North-Holland.

Raybould, B. (1998, June). Performance support engineering: An emerging development methodology for enabling organizational learning. Available online: www.arielpcs.com/barry4.htm.

Roth, E. M., Bennett, K. B., & Woods, D. D. (1987). Human interaction with an "intelligent" machine. *International Journal of Man-Machine Studies, 27*, 479-526.

Smeaton, A. (1994). Linguistic approaches to text management: An appraisal of progress. *Journal of Document and Text Management, 2*(2), 67-80.

Suchman, L. (1987). *Plans and situated actions: The problem of human-machine communication*. Cambridge, England: Cambridge University Press.

Terveen, L., Selfridge, P., & Long, D. (1995). Living design memory: Framework, implementation, lessons learned. *Human-Computer Interaction, 10*(1), 1-37.

Thuring, M., Hannemann, J., & Haake, J. (1995). Hypermedia and cognition: Designing for comprehension. *Communications of the ACM, 38*, 57-66.

Treu, S. (1990). Conceptual distance and interface-supported visualization of information objects and patterns. *Journal of Visual Languages and Computing, 1*, 369-388.

Tufte, E. (1983). *The visual display of quantitative information.* Cheshire, CT: Graphics Press.

Tversky, A., & Kahneman, D. (1981). The framing of decisions and the psychology of choice. *Science, 211*(30), 453-458.

Vincente, K. (1999). *Cognitive work analysis.* Mahway, NJ: Lawrence Erlbaum Associates.

Waern, Y. (1989). *Cognitive aspects of computer supported. tasks.* New York: Wiley.

Wickens, C. (1992). *Engineering psychology and human performance* (2nd ed.). New York: HarperCollins.

Winsor, D. (1996). Writing well as a form of social knowledge. In A. Duin & C. J. Hansen. (Eds.). *Nonacademic writing: Social theory and technology.* Mahwah, N J: Lawrence Erlbaum Associates.

Woods, D., & Roth. E. (1988). Cognitive engineering: Human problem solving with tools. *Human Factors, 30*(4), 415-430.

Yoon, W., & Hammer, J. (1988). Aiding operator during novel fault diagnosis. *IEEE Transactions on Systems, Man, and Cybernetics, 18*(1), 142-147.

11

Applying Survey Research Methods to Gather Customer Data and to Obtain User Feedback

Beverly B. Zimmerman
Maribeth C. Clarke
Brigham Young University

Documentation professionals frequently have to gather information about their users or obtain feedback from them. Much of this information gathering is based on a written or verbal question-and-answer process. Surveys can supplement other information-gathering methods such as focus groups, contextual inquiry, and case studies, and can be a cost-effective measure for helping documentation specialists obtain a broader sense of how users respond to products and documentation than they could develop from their own perspective.

For example, surveys can be used to do the following:

- Acquire demographic information about customers, including their education or training, their level of computer software expertise, and the specific hardware and printer systems they use,
- Gauge attitudes or opinions about software, upgrades, and potential new features,
- Provide documentation writers with an understanding of the context in which a particular software product is used,
- Obtain evaluative information about the company or software that can be communicated to other stakeholders.

In the past, documentation writers may have considered surveys to be too complex or too costly to be used for obtaining customer data, or they may have assumed that every survey required a trained statistician in order to be effective. Dillman (2000) pointed out, however, that these assumptions may no longer be correct as recent technological changes have expanded the opportunities for conducting surveys in less expensive ways and opened the possibility for people with limited research experience—such as documentation specialists and information designers—to do credible surveys.

At the same time, however, simply listing a few questions on a form to be mailed to customers or to be placed on a web site isn't enough to constitute an effective survey. Rose (1981) found that poorly written questionnaires were so overwhelming that many readers chose not to deal with them even when it meant giving up something they deserved and needed. Bagin and Rose (1991) reported that respondents complained about time-consuming surveys with complicated forms, unclear instructions, vague questions and requests for inappropriate information. The result was that these bad questionnaires and surveys cost companies untold dollars. MacNealy (1994) also found that poorly designed information-gathering forms drove up operating costs for businesses and other organizations. These results parallel those of Fisher (1999) who found users of online help complained about too technical language, too much superficial information, and inaccurate, incomplete, and out-of-date information. Faced with these problems, users had given up using the documentation, or worse still, had made costly mistakes in entering data.

Because a successful survey can provide valuable data for making decisions about software and documentation, it is important that designers of surveys understand how to carefully plan and implement an effective survey. The purpose of this chapter is to review recent work in survey research and cognitive psychology that will help documentation specialists understand what factors affect the quality and quantity of survey responses so that they can create effective and helpful surveys and to help them understand the cognitive processes involved in completing any difficult task or procedure.

The survey design process

Some of the earliest work on conducting effective surveys was done by Dillman (1978) who outlined a one-size-fits-all method for producing high response rates from mail surveys. Although Dillman's Total Design Method (TDM) was an improvement over survey practice at the time, its use of the same step-by-step protocol for every survey proved limiting.

New computer capabilities, evolving theories from cognitive science, and additional social science research, led Salant and Dillman (1994) to develop a more flexible method for designing surveys—one that acknowledged the varying needs of respondents, recognized the uniqueness of every survey situation, and implemented other survey modes, such as Internet surveys. Nevertheless, as Dillman (2000) recently argued, his original premise is still valid: Successful surveys require diligence and planning. Every aspect of a successful survey must be designed to obtain the best possible response, and the entire survey must be carefully administered and coordinated with a specific goal in mind.

Conducting a survey involves determining the characteristics, attitudes, or behaviors of a group or population by systematically interviewing a representative subgroup or sample of the population. Although every survey situation differs, the general process for designing a successful survey begins by first identifying a question or problem, determining what information must be obtained to answer the question, and then analyzing whether a survey is the best way to obtain the information. If a survey is appropriate, the process continues by selecting a survey method and establishing research objectives or specific goals. These objectives guide writers and designers in creating both the sampling methods for determining the survey sample and the survey instrument or questions for collecting the data. After the survey is implemented, data is collected, analyzed, and interpreted, and the results are used to answer the original question.

The survey process is composed of many constructs or variables that can affect the success of the overall survey. These constructs exist in every phase of the survey process and have been the object of study by every research discipline. For example, Christensen (1997) found research on the mailed questionnaire process that studied the application of science and technology to the questionnaire process; looked at the geographical, historical, political, psychological, and sociological aspects of the response process; observed the economic impact on the researcher's and respondent's resources, and observed the researcher's and respondent's abilities to communicate effectively.

Schwarz (1996) argued that in the past, survey research could be characterized as a rigorous approach toward sampling methodology and data collection on the one hand, and a lack of rigor in the design of the survey instrument on the other. Recently, however, work in social and cognitive psychology has lead to theoretical models on the cognitive and communicative processes that occur in surveys, and researchers have begun to test these models.

Documentation specialists are most likely to be involved in the question-asking/question-answering part of the survey process, that is, in creating and testing the survey instrument. However, they may be involved in every aspect of the survey process. In addition, much of the research on how users answer questions can be used to understand how people read and respond to any complicated document, including computer documentation. In this chapter, we focus on understanding factors affecting survey participation, namely the cognitive decision-making processes respondents use to complete a survey, factors influencing the design of a survey, and theories for increasing participation and response. We do not consider the processes for setting objectives, determining whether a survey is appropriate, or selecting a survey method (see Salant and Dillman [1994] for a nonacademic guide to these topics), nor do we address aspects of data analysis or the reporting of survey statistics.

Cognitive processes involved in completing a survey

Survey results are only as meaningful as the responses provided by participants. Unfortunately, documentation writers and designers may not appreciate the difficulty respondents have in answering even one simple question. Krosnick (1991) articulated the complex cognitive process respondents must go through in answering survey and other questions. First, respondents must "carefully interpret the meaning of each question, search their memories extensively for all relevant information, integrate that information carefully into summary judgments, and respond in ways that convey those judgments' meanings as clearly and precisely as possible" (p. 31).

When respondents perform each of these necessary cognitive tasks in an "effortful" manner to provide the best answer, they are said to be optimizing (Krosnick, 1991). However, when the task of completing a questionnaire becomes too demanding or time consuming, respondents may shift their response strategy to provide a reasonable or feasible answer rather than the optimal one. This behavior, which Krosnick calls *satisficing* (the term is borrowed from Simon, 1957), describes when respondents answer a question without exploring all the options. Weak satisficing occurs when respondents are less thorough in executing all of the steps of the process, such as selecting the first acceptable answer or agreeing with all of the true/false or yes/no questions. Strong satisficing occurs when respondents leave out altogether the retrieval and judgment steps of the response process, such as always selecting the status quo or no opinion options of a question (p. 32).

Krosnick, Narayan, and Smith (1996) identified three factors that increase the likelihood that a person responding to questions may begin to satisfice: (1) the greater the difficulty of the task; (2) the lower the ability of the person reading the questionnaire; and (3) the lower the person's motivation to optimize. Task difficulty is "likely to be a function of the complexity of the language" of the question, the respondent's familiarity with the concepts connected with the question, the respondent's process of retrieving information from memory, the complexity of the information the respondent must summarize in order to formulate an answer, and the ease with which the respondent's answer can be matched to the response alternatives that are offered (p. 33). Task difficulty increases when an interviewer reads the questions too quickly or the respondent experiences significant distraction while completing the survey.

Respondents differ in their cognitive ability as well as in their motivation to attend to a questionnaire, and this influences their response behavior. Krosnick et al. (1996) suggested that optimizing behavior is easier for respondents who are "adept at the process of retrieving information from memory" and expressing that information verbally, those who are experienced in thinking about the subject matter of a particular question, and those who have already formed an answer to the precise question that is asked. Presumably, respondents who are motivated to optimize would be those who value cognitive activities, find the topic of question to be personally important, feel the survey is valuable or useful, are moved by a request for "careful reporting," or feel "accountable to defend" their behaviors and opinions. The length of a survey or interview may also influence the behavior of respondents as increasing fatigue lowers their motivation to optimize (p. 32).

Interpreting the meaning and intent of a question

Understanding a survey question in order to formulate a meaningful answer, like understanding the directions for completing a task, involves more than simply comprehending the literal meaning of a sentence. Respondents must understand what the author of the question intended. Schwarz and Hippler (1991) asserted that respondents infer the intended meaning of a question based on their experience with the give-and-take nature of everyday conversation, a process which assumes that every utterance is truthful, informative, relevant, and clear (Grice, 1975). This notion would support Groves' (1989) findings that respondents will answer a question even when it is ambiguous or they do not understand it.

Survey respondents also use response alternatives to clarify the meaning of a question and to recall behaviors or opinions. Schwarz, Strack, Muller, and Chassein (1988) found respondents' answers were influenced by the frequency range of response alternatives as respondents used the specific numeric values provided in a rating scale to interpret the meaning of the scale's labels. In addition, respondents may draw on a preceding question to help them interpret a subsequent one. Schwarz & Hippler (1991) found respondents used preceding questions to determine the nature of their response even when the issue underlying the question was fictitious. Likewise, respondents exhibited *carryover*, reporting their satisfaction with life as a whole, based on a previous question asking them to report happiness with their marriage. Finally, respondents may interpret the intention of a question as it is influenced by the very context in which survey questions are asked, even going so far as to conjure up opinions to avoid appearing uninformed (Erikson, Luttbeg, & Tedin, 1988; Schuman & Presser, 1981).

Thus, respondents' assumptions about conversational cooperativeness influences their understanding of survey questions. As a result, Schwartz et al. (1996) argue there can be marked differences in answers to identically worded questions depending on whether or not response alternatives are provided, by the specific response alternatives that are provided, and by the order in which questions are asked. These *response effects* draw respondents' attention to information they might not have considered or influence how they interpret and answer survey questions.

Recalling and integrating information

After respondents have interpreted the meaning and intent of a survey question, they must recall relevant information from their memory and integrate that information into a single answer. Many times, however, respondents cannot readily call up an appropriate answer, especially if the question asks about their behavior. Research on the recall of behavior (Linton, 1982; Neisser 1986; Strube, 1987; Schwarz & Sudman, 1994) reveals that respondents blend the details of various instances of a particular behavior into one overall experience. As a result, respondents tend to overestimate the occurrence of events that occur regularly and underestimate the occurrence of events that occur irregularly (Fiedler & Armbruster, 1994). Respondents also use their current behavior or opinion as a benchmark to underestimate the degree of change over time (Ross, 1989), use response alternatives as a frame to limit the range of their behavioral occurrences (Schwarz & Hippler, 1991), and adopt a survey's vague quantifiers (such as "sometimes," and "frequently") to denote an objective frequency (Pepper, 1981).

Survey designers have attempted to simplify the respondents' task in reporting behavior by providing recall cues or by restricting recall to a specific time period. But these strategies have drawbacks because respondents may restrict their memory search and omit valid occurrences simply because they differ from the specified cues or time periods

(Schwartz et al., 1996). Wagenaar (1986) found that the poorest recall cue is the date of an event. More effective cues involve the so-called reporter's questions—the what, where, and who of an event. Providing respondents with adequate time to search memory also improves recall (Reiser, Black, & Abelson, 1985), as does asking respondents to start with the most recent occurrence of a behavior and to search backward in time (Loftus & Fathi, 1985; Whitten & Leonard, 1981).

Survey questions regarding respondents' attitudes pose a different set of problems. Schwartz et al. (1996) note that whereas behavior can be measured in a more straightforward fashion, attitudes are subjective and dependent on context; that is, they exist relative to the situation in which the survey attempts to measure them. Efforts to create clear and unbiased questions can be difficult because surveys by their very nature can only address complex social issues (such as gun control) in a limited manner and results might differ if another facet of the issue were addressed or emphasized. Thus, Schwarz et al. (1996) warn users of survey results not to assume that responses to one aspect of an issue are representative of attitudes toward the issue in general. In addition, apparently minor changes in the wording of a question can lead to changes in responses (Bradburn, 1983). For example, introducing a reference to "communist" activities increased support for military operations (Schuman & Presser, 1981). On the other hand, efforts to avoid commonly used "loaded" terms may influence respondents' to adjust their real attitudes toward an issue.

Other, less substantive, changes in wording may change the nature of the respondents cognitive task. For example, Schuman and Presser (1981) replicated what is known as the forbid–allow asymmetry: Respondents more often will agree that something should not be forbidden than that it should be allowed and more often will agree that something should not be allowed than that it should be forbidden. Hippler and Schwarz (1986) concluded that respondents respond "no" to both questions because they adjust their task depending on the form of the question and focus on the implications of forbidding or allowing something, rather than on the implications of not doing it.

In addition, respondents tend to accept the constraints imposed by a question. For example, respondents are more likely to report not having an opinion when a "don't know" response is listed (Schwarz & Hippler, 1991). Schuman and Scott (1987) provided respondents with an open-ended question asking them to list what they believed to be the most important events of the 1950s, and only a few listed the invention of computers. However, when the invention of computers was listed as a response alternative in a closed question, respondents selected that response most frequently.

Providing an appropriate response

After interpreting the meaning of a question and recalling relevant information, respondents must now respond in terms of the response alternatives provided in the questionnaire. Jansen and Steehouder (1992) developed a model of different subtasks that appear to be necessary for respondents to provide an appropriate response to the questions on a survey form. These include functional tasks (generating an item of information, verifying whether that information fits the conditions of the survey question, and transforming the information into a verbal answer or selecting an appropriate response on a form), interpretation tasks (understanding the semantic aspects of the text at both a local and general level and determining how to perform the functional subtasks), and monitoring tasks (orienting, selecting, and checking the form). The researchers concluded that problems can occur in each type of subtask. For example, respondents may have difficulty in getting an overview of the form and the activities they must perform, understanding textual and graphical instructions, interpreting the syntactic structure and terminology of questions, and performing calculations that may be required in order to reply to specific questions.

It is evident from the earlier discussion that obtaining answers to survey questions, just like any procedural task, is always problematic; nevertheless, designers and writers can do much to decrease the difficulty of the response task by understanding how the cognitive nature of the task is influenced by other factors and using this understanding to develop the best questionnaire or set of instructions possible.

Creating the survey instrument

Developing appropriate questions and creating the survey instrument is the most critical aspect of a survey, yet it often is the least emphasized. As pointed out earlier, in comparison to the vast amount of research on aspects of sampling, only a limited amount of research has been done on creating effective survey instruments and questions. What information is available usually takes the form of guidelines for what is believed to be effective, rather than results of experimental research. For example, Dillman (2000) provided excellent guidelines on writing questions, constructing a questionnaire, and merging the verbal and graphical language of the questionnaire based on his revised TDM model, substituting a "tailored" approach for his previous one-size-fits-all approach.

Some important research has been completed, however, and research in this area is beginning to grow. For example, Wright (1981) determined that checking boxes was faster and more effective than underlining a correct answer or deleting an incorrect one, and that ticking adjectives (e.g., single/married) was faster and more effective than replying to yes/no questions. Holland and Redish (1981) found respondents who completed forms properly paid more attention to the rhetorical situation or overall purpose of the form as well to the function of individual questions, whereas less successful respondents concentrated on simply decoding each question.

Jansen and Steehouder (1992) noted substantial discrepancies between what designers of survey questionnaires believed were the attitudes, knowledge, and skills of respondents and what the respondents actually exhibited. These results should be of special interest for documentation writers. For example, respondents did not systematically read through explanatory information, nor did they follow instructions as designers anticipated. In addition, respondents lacked the reading proficiency to understand explanatory instructions; failed to understand graphical signposts such as arrows, colors, and typefaces; and lacked arithmetic skills to complete simple calculations. Respondents also lacked the background knowledge necessary to complete the form successfully, failed to understand the purpose for the survey, and were unclear on the importance of particular information.

On the basis of their research, Jansen and Steehouder outlined specific criteria for creating survey forms. First, questions should be written from the perspective of the specific situation or task being posed and should be based on the *scenario principle* (Flower, Hayes, & Swartz, 1983), that is, written in the active voice using action verb sentences with people or organizations as the doer of the action and the subject of the sentence. Second, writers should make clear what is expected of readers by providing them with clear instructions for completing the survey, including help in solving any problems in completing tasks, such as "routing instructions" explaining how to skip irrelevant questions. Third, designers should provide adequate background information such as the purpose of the survey in order to compensate for the respondent's lack of experience with the form. Ideally, this statement should be easily identifiable by the respondent and should be read prior to completing the form. However, because few users actually read these summaries, writers and designers should compensate by also including relevant instruction and information alongside the questions to which they refer.

Conclusions drawn from survey research parallels that done in technical communication by Isakson and Spyridakis (1999) who concluded that readers who are familiar with and interested in a topic read a text more easily than those who are less familiar and less interested and suggested that writers provide background information, employ analogies and metaphors, and add illustrations and explanations.

Other survey research shows the value of supplementing traditional pilot testing of surveys using cognitive methods that provide insight into how respondents respond to the form. These methods can help designers uncover many problems and enable them to revise the form. For example, Graesser, Bommareddy, Swamer, and Golding. (1997) used a cognitive computation model to revise survey questions and concluded that revised questions produced more reliable answers than the original ones. Zimmerman and Schultz (2000) implemented evaluation by experts and user feedback-based usability testing to create a medical questionnaire that significantly outperformed a form created using the default settings of a survey software program.

Presser and Blair (1994) compared four survey pretest methods (informal field testing, behavior coding, cognitive interviews, and panels of experts). They found expert panels to be the least expensive method and they also identified the most number of problems. Expert panels, along with cognitive interviewing, were most effective in identifying specific types of problems and conventional pretesting and behavior coding were best at identifying problems involving an interviewer. Presser and Blair concluded, however, that more research needs to be done on how to help survey designers discover substantial systematic errors by respondents.

Motivating respondents to respond

A predominant problem with using surveys to gather data has been motivating respondents to respond. When large numbers of individuals in the sample are unwilling or unable to participate in a survey, the survey itself may be compromised because there is no guarantee that the results obtained from the sample are representative of the entire population. For mailed surveys, a 50% response rate is considered adequate and a 70% response is considered very good; however, the typical return rate is around 30%, especially in surveys of the general public (Leedy, 1993). Face-to-face surveys seldom achieve response rates higher than 70%, and telephone surveys have difficulty achieving response rates higher than 60% (Brehm, 1993). Groves and Kahn (1979) found that people prefer some survey modes over others and there is a growing trend to employ multiple survey methods in the same survey. This trend, however, creates a paradox. Higher response rates can usually be achieved using more than one survey mode, but researchers have found differences in the answers respondents provide to different survey methods (de Leeuw, 1992; Schwarz, Hippler, & Noelle-Neumann, 1992).

Low response rates are thought to be discriminatory of individuals who lack interest in the topic of the survey, are extremely busy or live in big cities, differ culturally from the norm of the sample, have high income levels, or have literacy problems, vision problems, or psychomotor impairment (Couper & Groves, 1996). Additionally, low response rates can be either overly representative or discriminatory of low educational backgrounds (Brehm, 1993; Leedy, 1993; Shaughnessy & Zechmeister, 1994; Smith, 1983). Chou (1997) found online surveys discourage some persons from completing the survey because they were not familiar with the computer technology used.

Recently, the notion that high response rates are critical for representativeness in survey research has been challenged. Krosnick (1999) argued that increasing response rates does not necessarily increase representativeness. He cites, for example, Greenwald, Carnot, Beach, and Young (1987) who conducted telephone surveys of the general public just before an election and then checked voting records to see whether each respondent voted. Results showed that the harder it was to contact persons, the less likely it was that they voted. Thus Couper and Groves (1996) conclude that working to boost the response rate, in this instance, would actually decrease the representativeness of the sample. Visser, Krosnick, Marquette, and Curtin (1996) compared survey forecasting to actual election results and found mail survey response rates of 20% were more accurate in predicting election results (average error = 1.6%) than were telephone surveys with 60% response rates (average error = 5.2%). The mail surveys were also more representative of voter demographics. Nevertheless, although it may no longer be presumed that higher response rates necessarily increase representativeness, writers and designers should use careful planning and sound methodology to prevent low response rates and to insure that their survey is a viable tool for gathering customer data.

The determinants of response

Answering the question, "What determines whether a person will complete a questionnaire?" is complicated. Couper and Groves (1996) proposed a model of survey participation that separated four sets of influences affecting survey cooperation into two categories: those that are out of the researcher's control and those that are under the researcher's control. The respondent's social environment and household(er) factors such as sociodemographic characteristics and psychological predisposition were considered to be outside the control of the survey designer. Factors under the researcher's control included the design of the survey and the characteristics of interviewers. However, the model shows that an interaction occurs between the respondent and the survey and/or the interviewer. Factors that the respondent deems most relevant to the decision to participate form the basis upon which the respondent decides to participate or refuses to participate in the survey request. Thus, the ultimate decision by a customer to participate in a survey is based on the combined influence of interacting constructs, some facilitating cooperation and others constraining it.

Christensen (1997) reported 12 main determinants that are believed to influence the response process in varying ways:

1. Physiological barriers.
2. Ability to be accurate.
3. Known benefits of response.
4. Known costs.
5. Perceived salience (importance).
6. Opportunity costs of response.
7. Financial rewards.
8. Financial costs.
9. Immediacy (urgency) restraints.
10. Intellectual ability to comprehend and respond.
11. Philosophical commitment to the response process.
12. Aesthetic stimulus of the questionnaire.

Designers of surveys can understand and try to indirectly influence these factors by asking the question, "What opportunity does the respondent give up to actually read and complete the questionnaire?" However, these factors are difficult to analyze—even though they may be significant in the respondent's decision to respond—and may remain unknown to the designer. These factors can also differ on any given day and at any given time.

The various determinants of response impact the reply process in numerous ways and to varying degrees. Thus, very slight variances can amplify into unpredictable results and a small change can have an impact on the response far beyond what could have been predicted. Although the hope exists to gain predictability of the questionnaire response process, no level of detail will ever satisfy this desire completely (Briggs & Peat, 1989; Coveney & Highfield, 1990; Garcia, 1991; Gleick, 1987; Wheatley, 1993). It is assumed, however, that some determinants of response are more significant in the overall response rate achieved from a questionnaire. The topic of which determinants are significant is discussed in a later section.

Response inducements

For three decades, researchers have generated a plethora of research on the impact of various constructs on mailed questionnaire return. More than 90% of the research on survey response relates to response inducement techniques (methods for triggering or enhancing response) that can be implemented in the process of constructing and administering the questionnaire (Boser & Clark, 1993; Fox, Crask, & Kim, 1988; Hopkins & Gullickson, 1989; Rodgers & Worthen, 1995; Yammarino, Skinner, & Childers, 1991), while the other 10% focused on theoretical concepts to explain the behavior motivation required to induce response (Biner, 1988; Dillman, 1978, 1991; Furse & Stewart, 1982; Hantula, Stillman, & Warnach, 1990; Heberlein & Baumgartner, 1978; Lockhart, 1984; McKillips, 1984).

There have been at least 16 comprehensive reviews or meta-analyses attempting to clarify the significant response inducement variables of mailed questionnaires resulting from the 300 or more studies on response inducement techniques. The methodology used within these reviews range from descriptive (Kanuk & Berenson, 1975; Linksy, 1975) to narrative (Conant, Smart, & Walker, 1990; Duncan, 1979; Harvey, 1987) to meta-analytic (Fox et al., 1988; Heberlein & Baumgartner, 1978; Yammarino et al., 1991). Although these meta-analyses differ in the year they were published and the number of primary studies analyzed, an analytic tally of the results of these reviews illustrates that the use of follow-up letters, monetary incentives, prenotification, sponsorship of the survey, and the type of postage used are the methods most effectively employed to increase questionnaire response rates. In a meta-analysis of nine of these comprehensive reviews, Rodgers and Worthen (1995) confirmed the use of follow-up letters, monetary incentives, and prenotification as being the most effective response inducement techniques and questioned the other variables such as sponsorship, type of postage used, and the length of questionnaire, which are often cited by other researchers as being significant.

Although the increases in response rates attributable to specific response inducement techniques are important, it is even more important to examine the total percentage of response achieved in any given survey research study. Christensen (1997) analyzed 298 studies utilizing a response inducement technique to enhance questionnaire return and found the overall range of response rates achieved in those studies varied dramatically from 40% to 80%. Additionally, the use of a response inducement technique to improve response rates only accounted for a mean of a 6% to 20% increase in a survey's response rate. When the percentage rates of return were averaged for each of the response inducements, the results showed that the geographic distribution of the sample, the use of follow-up reminders, prenotification, and questionnaire format and length had the greatest impact on increasing or decreasing the percentage of response.

If the increase in response rates is attributed solely to the introduction of specific response inducements (each of which typically provide 5% to 10% increases), then a mailed questionnaire survey employing 10 to 20 treatments could, hypothetically, have a response rate above 100%, if the treatments were additive rather than interactive. An alternative assumption would be that a synthesis of the multiple variables produces a ceiling effect, but this doesn't resolve the question of why the same questionnaire survey, using identical methodology, mailed to two different populations produces an entirely different response rate. Therefore, the vast range of response rates achieved should indicate that the introduction of a response inducement technique is not the only or the most significant determinant of response.

Christensen's (1997) synthesis of research on high versus low response rates, percent of responses, and response inducement techniques showed that 11 response variables appear to play a more significant role in mailed questionnaire surveys response rates then do other inducement variables. These include personalization of the cover letter, signature sponsorship of the cover letter, level of threat posed by the questionnaire, geographical location of the sample, type of postage used, where the questionnaire was received, questionnaire length, questionnaire format, follow-up reminder notices, prenotification, and the awarding of monetary incentives. From a theoretical perspective, the first five response inducement techniques listed are part of the perceived salience (importance) of response and the last six response inducement techniques listed are part of the perceived immediacy (urgency) of response.

Salience and immediacy as significant determinants of response

Salience (the quality of being important) motivates respondents to act upon a survey. Respondents may be motivated by the survey's geographical salience (the composition, size, and location of the sample), its political salience (the topic and the sponsor of the questionnaire), its psychological salience (resulting from the communication to the respondent in the letter of transmittal, particularly in how it conveys a level of threat, anonymity, confidentiality, personalization, appeal for response, and expressed value of the response and the respondent) or its sociological salience (how the respondent perceives the social desirability of the research topic and objectives). If salience is judged as high, then salience becomes a strong determinant in the respondent's decision to complete and return a survey questionnaire.

Immediacy (the quality of being urgent) prompts respondents to take direct action and provides them with freedom from the feeling of need for immediate intervention. Immediacy facilitates the multitude of temporal management responsibilities a respondent faces. Immediacy is economically motivated by providing freedom from external constraints that impact the allocation of time, the management of resources which may constrain energy and intellect, and/or through the reward of money to compensate for the use of these resources. When immediacy is involved in the decision to respond to a mailed questionnaire it will dictate the urgency and speed of the respondents' efforts. Immediacy improves the level of response when it is coupled with salience (Baumgartner & Heberlein, 1984; Jobber, 1984) and by creating urgency appears to be an even stronger motivator for immediate response than salience (Christensen, 1997). Thus, the survey designer or documentation writer who strives to enhance and control salience and immediacy in the response process will achieve a higher response rate and a more representative sample of completed questionnaires.

Motivating users to respond

A few researchers have attempted to explain the motivation of the respondent based on the idea that a customer responds to a questionnaire because of its importance to them or that the importance of the questionnaire is enhanced by an urgency to respond. For example, Heberlein and Baumgartner (1978) and Jobber (1984) all hypothesized that the salience of a survey topic is a high predictor of response rates. McKillips' (1984) Attitude Theory argued that the importance of the survey is tied to a respondent's personal value system. The Opportunity Cost Hypothesis (Couper & Groves, 1996), is similar in thought and asserts that questionnaire response is weighted in the judgment of what opportunity the respondent will give up in order to respond to a questionnaire. Biner's (1988) Reactance Theory argued that when the importance of the research and the urgency with which one should respond are stressed, the result may be an inadvertent threat to the respondent's freedom. Freedom is returned only after the respondent returns the questionnaire. None of these theories, however, integrates the ideas of both salience and immediacy as they impact survey respondents.

Figure 11.1 presents Christensen's (1997) model of the interaction of immediacy and salience in the mailed questionnaire response process. In the model, a four-quadrant format juxtaposes salience and immediacy against each other. Each of the four quadrants predicts the expected level

of response generated from the interaction of the two variables. Immediacy is placed on the horizontal axis of the quadrants while salience is
placed on the vertical axis of the quadrants. The arrows on the model indicate
that there is a continuum within and between each of the four quadrants
because the quadrants are not truly linear and may overlap at times.

In Quadrant I of the model, responding to the questionnaire is both
salient and immediate to the respondent and the respondent is motivated to return the questionnaire. In principle, the questionnaire "acts
on" the respondent (Polit & Hungler, 1999). With both immediacy and
salience at high levels, this quadrant predicts high response. In Quadrant
II, responding to the questionnaire is salient, but not immediate. The
respondent is not motivated by the questionnaire itself, but is motivated
to complete the questionnaire by the salience of the topic, sponsor, or
researcher. In essence, the respondent "acts on" the questionnaire and
decides to complete the questionnaire because it is deemed important
(salient). Because of the salience of the topic, the sponsor, or the researcher, the returned questionnaire is likely to be completed thoroughly
and correctly, which decreases response bias. Nevertheless, without high
immediacy, the overall response rate in this quadrant is hypothesized to
be only moderate.

In Quadrant III, the completion of the questionnaire is immediate
but not truly salient to the respondent. The sense of immediacy stems

HIGH IMMEDIACY ▶	◀ NO IMMEDIACY
I EXTREMELY HIGH RESPONSE	**II** MODERATELY HIGH RESPONSE
Responding to the Questionnaire Is Immediately Important	Questionnaire Will Be Completed Between Other Pressing Demands, if Possible
III MODERATE RESPONSE	**IV** LOW RESPONSE
Questionnaire Will Be Completed if Convenient	Responding to the Questionnaire Is not Immediately Important

(Left axis: HIGH ◀ SALIENCY for top rows; NO SALIENCY ▶ for bottom rows)

Figure 11.1. Model for the interaction of immediacy and salience in
questionnaire response.

from the illusion of salience that likely exists for someone besides the respondent, for example, the researcher or sponsor. Respondents complete and return the questionnaire more out of a need to remove it from their realm rather than as a personal desire to do so. Questionnaires imbedded in this quadrant will achieve only a moderate level of response. Designers and writers should not mistake Quadrant I with Quadrant III as they plan, develop, and then pretest questionnaires. Although the writer may believe that the questionnaire has high salience, only the respondent determines if the questionnaire actually has salience. The same problem could be encountered between Quadrant I and II if the writer misjudges what is immediate to a group of respondents.

Quadrant IV has the lowest predicted response rate, which comes as a result of a questionnaire that has no immediacy or salience to the respondent. Questionnaire response depends on the willingness of the respondent to take the time to complete the questionnaire rather than to throw it away (Covey, Merrill, & Merrill, 1994; Christensen, 1997).

A model such as this, corresponds to research in technical communication (Isakson & Spyridakis, 1999) concluding that readers recall information they perceive to be important easier than information that they find to be unimportant. Successful use of the model, however, is dependent upon a writer's ability to determine what constitutes saliency and immediacy with the intended audience and to integrate those factors into the survey or other written documents.

Summary

Surveys can provide valuable information for making decisions about software and documentation. Surveys can be used to acquire demographic information about customers, including their education or training, their level of computer software expertise, and the specific hardware and printer systems they use. Surveys can be used to gauge attitudes or opinions about software, upgrades, and potential new features. Surveys can be used to provide documentation writers with an understanding of the context in which a particular software product is used. And surveys can be used to obtain evaluative information about the company or software that can be communicated to other stakeholders such as software developers, shareholders, or management.

Recent work in survey research and cognitive psychology shows that respondents go through a complex cognitive process in answering even one simple survey question. Designers and writers can do much to increase the likelihood that a person will respond to survey questions in an optimal manner by decreasing the difficulty of the task and increasing the respondent's motivation to optimize. In addition, respondents can be motivated to complete and return surveys by how the designer uses response inducement techniques and by the degree to which the respondent perceives the salience and immediacy of the survey. Conclusions from survey research parallel conclusions drawn from research in technical communication, namely that readers' perceptions about the text influence their performance. Readers who are interested with a topic and find it important read a text more easily than those who are not.

References

Bagin, C. B., & Rose, A. M. (1991). Worst forms unearthed: Deciphering bureaucratic gobbledygook. *Modern Maturity, 34,* 64-66.

Baumgartner, R. M., & Heberlein, T. A. (1984). Recent research on mailed questionnaire response rates. *New Directions for Program Evaluation, 21,* 65-76.

Biner, P. M. (1988). Effects of cover letter appeal and monetary incentives on survey response: A Reactance Theory application. *Basic and Applied Social Psychology, 9*(2), 99-106.

Boser, J. A., & Clark, S. B. (1993, April). *Response rates in mail surveys: A review of the reviews.* Paper presented at the annual meeting of the American Educational Research Association, Atlanta, GA.

Bradburn, N. M. (1983). Response effects. In P. H. Rossi & J. D. Wright (Eds.), *The handbook of survey research.* New York: Academic Press.

Brehm, J. (1993). *The phantom respondents.* Ann Arbor: University of Michigan Press,

Briggs, J., & Peat, F. D. (1989). *Turbulent mirror: An illustrated guide to chaos theory and the science of wholeness.* New York: Harper & Row.

Chou, C. (1997). Computer networks in communication survey research. *IEEE Transactions on Professional Communication, 40,* 197-207.

Christensen, M. (1997). An interdisciplinary theoretical framework for the mailed questionnaire process and the development of a theory on immediacy and salience as significant variables of response rates (Doctoral dissertation, Utah State University, 1996). *Dissertation Abstracts International: Section B: The Sciences & Engineering, 57*(7-B, 47-79.

Conant, J. S., Smart, D. T., & Walker, B. J. (1990). Mail facilitation techniques: An assessment and proposal regarding reporting practices. *Journal of the Market Research Society, 32*(4), 569-579.

Couper, M. P., & Groves, R. M. (1996). Household-level determinants of survey nonresponse. In M. T. Braverman (Ed.), *Advances in survey research.* New directions for program evaluation (no. 70, pp. 63-70). San Francisco: Jossey-Bass.

Coveney, P., & Highfield, R. (1990). *The arrow of time: A voyage through science to solve time's greatest mystery.* New York: Fawcett Columbine.

Covey, S. R., Merrill, A. R., & Merrill, R. R. (1994). *First things first.* New York: Simon & Schuster.

de Leeuw, E. D. (1992). *Data quality in mail, telephone and face-to-face surveys.* Unpublished dissertation, Vrije Universiteit, Amsterdam.

Dillman, D. A. (1978). *Mail and telephone surveys: The Total Design Method.* New York: Wiley.

Dillman, D. A. (1991). The design and administration of mail surveys. *Annual Review of Sociology, 17,* 225-249.

Dillman, D. A. (2000). *Mail and Internet surveys: The Tailored Design Method.* New York: Wiley.

Duncan, W. J. (1979). Mail questionnaires in survey research: A review of response inducement. *Journal of Management, 5*(1), 39-55.

Erikson, K. A., Luttbeg, N. R., & Tedin, K. L. (1988). *American public opinion.* New York: Macmillan.

Fiedler, K., & Armbruster, T. (1994). Two halfs may be more than one whole: Category-split effects on frequency illusions. *Journal of Personality and Social Psychology, 66,* 633-645.

Fisher, J. (1999). Technical communicators and online Help: The developers' and users' perspectives. *Technical Communication, 46*(3), 360-381.

Fox, R. J., Crask, M. R., & Kim, J. (1988). Mail survey response rate: A meta-analysis of selected techniques for inducing response. *Public Opinion Quarterly, 52*(4), 467-491.

Flower, L., Hayes, J., & Swarts, H. (1983). Revising functional documents: The scenario principle. In P. Anderson, R. Brockman, & C. Miller (Eds.), *New essays in technical and scientific communication: Research, theory & practice,* (pp. 41-58). Amityville, NY: Baywood.

Furse, D. J., & Stewart, D. W. (1982). Monetary incentives versus promised contribution to charity: New evidence on mail survey response. *Journal of Marketing Research, 19,* 375-380.

Garcia, L. (1991). *The fractal explorer.* Santa Cruz, CA: Dynamic Press.

Gleick, J. (1987). *Chaos: Making a new science.* New York: Viking.

Graesser, A. C., Bommareddy, S., Swamer, S., & Golding, J. M. (1997). Integrating questionnaire design with a cognitive computational model of human question answering. In N. Schwarz & S. Seymour (Eds.), *Answering questions: Methodology for determining cognitive and communicative processes in survey research* (pp. 143-174). San Francisco: Jossey-Bass.

Greenwald, A. G., Carnot, C. G., Beach, R., & Young, B. (1987). Increasing voting behavior by asking people if they expect to vote. *Journal of Applied Psychology, 72,* 315-318.

Grice, H. P. (1975). Logic and conversation. In P. cole & J. L. Morgan (eds). *Syntax and semantics, Vol. 3: Speech acts* (pp. 41-58). New York: Academic Press.

Groves, R. M. (1989). *Survey errors and survey costs.* New York: Wiley.

Groves, R. M., & Kahn, R.L. (1979). *Surveys by telephone* . San Diego: Academic Press.

Hantula, D. A., Stillman, F. A., & Warnach, H. R. (1990). A comparison of strategies for facilitating smoking survey return in low-SES employees. *Journal of Organizational Behavior Management, 11*, 47-59.

Harvey, L. (1987). A research note on the impact of class-of-mail on response rates to mailed questionnaires. *Journal of the Market Research Society, 28*, 299-300.

Heberlein, T. A., & Baumgartner, R. (1978). Factors affecting response rates to mailed questionnaires: A quantitative analysis of the published literature. *American Sociological Review, 43*, 447-462.

Hippler, H. J., & Schwarz, N. (1986). Not forbidding isn't allowing: The cognitive basis of the forbid-allow asymmetry. *Public Opinion Quarterly, 50*, 87-96.

Holland, V. M., & Redish, J. C. (1981). *Strategies for understanding forms and other public documents*. Document Design Center, American Institutes for Research, Washington, D.C.

Hopkins, K. D., & Gullickson, A. R. (1989). *Monetary gratuities in survey research: A meta-analysis of their effects on response rates*. Paper presented at the Annual Meeting of the American Educational Research Association, San Francisco, CA. (ERIC Document Reproduction Service No. ED 308 224).

Isakson, C. S., & Spyridakis, J. H. (1999). The influence of semantics and syntax on what readers remember. *Technical Communication, 46*(3), 366-381.

Jansen, C., & Steehouder, M. (1992). Forms as a source of communication problems. *Technical Writing and Communication, 22*(2), 179-194.

Jobber, D. (1984). Response bias in mail survey: Further evidence. *Psychological Reports, 54*, 891-894.

Kanuk, L., & Berenson, C. (1975). Mail surveys and response rates: A literature review. *Journal of Marketing Research, 12*, 440-453.

Krosnick, J. A. (1991). Response strategies for coping with the cognitive demands of attitude measures in surveys, *Applied Cognitive Psychology, 5*, 213-236.

Krosnick, J. A. (1999). Survey Research. *Annual Review of Psychology, 537*.

Krosnick, J. A., Narayan S., & Smith, W. R. (1996). Satisficing in surveys: Initial evidence. In M. T. Braverman (Ed.), *Advances in survey research. New directions for program evaluation* (no. 70, pp. 29-44). San Francisco: Jossey-Bass.

Leedy, P. D. (1993). *Practical research: Planning and design* (5th ed.). New York: Macmillan.

Linsky, A. S. (1975). Stimulating response to mailed questionnaires: A review. *Public Opinion Quarterly, 39*, 82-101.

Linton, M. (1982). Transformations of memory in everyday life. In U. Neisser (Ed.), *Memory observed: Remembering in natural contexts* (pp. 77-91). San Francisco: Freeman.

Loftus, E. F., & Fathi, D. C. (1985). Retrieving multiple autobiographical memories. *Social Cognition, 3,* 280-295.

Lockhart, D. C. (1984). The stages of mailed questionnaire returning behavior. *New Directions for Program Evaluation, 21,* 89-98.

MacNealy, M. S. (1994). Designing information-gathering forms. *Proceedings of the STC Annual Conference.* Arlington, VA: Society for Technical Communication, 440-442.

McKillips, J. (1984). Applying attitude theories to the return of mailed questionnaires. *New Directions for Program Evaluation, 21,* 77-87.

Neisser, U. (1986). Nested structure in autobiographical memory. In D. C. Rubin (Ed.). *Autobiographical memory* (pp. 71-88). Cambridge, England: Cambridge University Press.

Pepper, S. C. (1981). Problems in the quantification of frequency expressions. In D. W. Fiske (Ed.), *Problems with language imprecision* (New Directions for Methodology of Social and Behavioral Science, Vol. 9). San Francisco: Jossey-Bass.

Polit, D., & Hungler, B. P. (1999). *Nursing research: Principles and methods* (6th ed.). Philadelphia: Lippincott Williams & Wilkins.

Presser, S., & Blair, J. (1994). Survey pretesting: Do different methods produce different results? In P. Marsden (Ed.), *Sociological methodology 1994.* Oxford: Blackwell.

Reiser, B. J., Black, J. B., & Abelson, R. P. (1985). Knowledge structure in the organization and retrieval of autobiographical memories. *Cognitive Psychology, 17,* 89-137.

Rodgers, P. L., & Worthen, B. R. (1995, April). *A meta-analysis of factors that influence the response rate of mailed questionnaires.* Paper presented at the Annual Meeting of the American Educational Research Association, San Francisco, CA.

Rose, A. M. (1981). Problems in public documents. *Information Design Journal, 2,* 179-196.

Ross, M. (1989). The relation of implicit theories to the construction of personal histories. *Psychological Review, 96,* 341-157.

Salant, P., & Dillman, D. A. (1994). *How to conduct your own survey.* New York: Wiley.

Schuman, H., & Presser, S. (1981). *Questions and answers in attitude surveys.* New York: Academic Press.

Schuman, H., & Scott, J. (1987). Problems in the use of survey questions to measure public opinion. *Science, 236,* 957-959.

Schwarz, N. (1996). Survey research: Collecting data by asking questions. In G. R. Semin & K. Fielder (Eds.) , (pp. 65-90). *Applied social psychology.* Thousand Oaks, CA: Sage.

Schwarz, N., Groves, R.M., & Schuman, H. (1996). *Survey methods: The handbook of social psychology* (4th ed., Vol. 1, pp. 143-179), New York: McGraw-Hill.

Schwarz, N., & Hippler, H. J. (1991). Response alternatives: The impact of their choice and ordering. In P. Biemer, R. Groves, N. Mathiowetz, & S. Sudman (Eds.), *Measurement error in surveys* (pp. 41-56). Chichester, England: Wiley.

Schwarz, N., Hippler, H.J., & Noelle-Neumann, E. (1992). A cognitive model of response-order effects in survey measurement. In N. Schwarz & S. Sudman (Eds.), *Context effects in social and psychological research*, New York: Springer-Verlag.

Schwarz, N., Strack, F., Muller, G., & Chassein., B. (1988). The range of response alternatives may determine the meaning of the question: Further evidence on informative functions of response alternatives. *Social Cognition, 6*, 107-117.

Schwarz, N., & Sudman, S. (1994). *Autobiographical memory and the validity of retrospective reports*. New York: Springer-Verlag.

Shaughnessy, J. J., & Zechmeister, E. B. (1994). *Research methods in psychology* (3rd ed.). New York: McGraw-Hill.

Simon, H. A. (1957). *Models of man*. New York: Wiley.

Smith, T. W. (1983). The hidden 25 percent: An analysis of non-response in the 1980 General Social Survey, *Public Opinion Quarterly, 45*, 386-404.

Strube, G. (1987). Answering survey questions: The role of memory. In H. J. Hippler, N. Schwarz, & Sudman, S. (Eds.), *Social information processing and survey methodology* (pp. 86-101). New York: Springer-Verlag.

Visser, T. S., Krosnick, J. A., Marquette, J., & Curtin, M. (1996). Mail surveys for election forecasting? An evaluation of the Columbus Dispatch poll, *Public Opinion Quarterly, 60*, 181-227.

Wagenaar, W. A. (1986). My memory: A study of autobiographical memory over six years. *Cognitive Psychology, 18*, 225-252.

Wheatley, M. A. (1993). Comprehending chaos. *Brigham Young Magazine, 2*, 21-27.

Whitten, W. B., II, & Leonard, J. M. (1981). Directed search through autobiographical memory. *Memory and Cognition, 9*, 566-579.

Wright, P. (1981). Informed design for forms. In R. Easterby & H. Zwaga (Eds.), *Information design: The design and evaluation of signs and printed material* (pp. 545-577) Chichester, England.

Yammarino, F. J., Skinner, S. J., & Childers, T. L. (1991). Understanding mail survey response behavior: A meta-analysis. *Public Opinion Quarterly, 55*, 613-639.

Zimmerman, B. B., & Schultz, J. R. (2000). A study of the effectiveness of information design principles applied to clinical research questionnaires. *Technical Communication, 47*(2), 177-194.

12

Single Sourcing and Information Design

Ann Rockley
The Rockley Group, Inc.

Information developers have begun to create content that is designed to meet the specific needs of the user, the materials to be developed (such as user guides, reference guides, and training), and the delivery media (paper or online). Information design is critical to successful single sourcing. Creating effective content for single sourcing through information databases and in support of dynamic documents requires highly structured content based on information models. This chapter focuses on information design models for effective single sourcing.

What is single sourcing?

Traditional documents are written in files that consist of chapters or sections. These files are assembled to create a document. Help topics are also contained in files, which are often grouped by section. Single source materials are materials that come from a single location and consist of information objects, not files. The information objects are assembled to create information products. Therefore, information is written once, eliminating redundancy and errors that can occur when there is more than one source of content.

Typically, single sourcing requires that content is contained in a database or content management system. Single source information is broken down to the element level (section, paragraph, sentence). With this level of granularity, it is easy to select an element to reuse or repurpose it. There are three levels of single sourcing which will be discussed in the next sections.

Level 1: Identical content, multiple media

For a number of years, many writers have been creating Level 1 single source materials when they produced Help from their paper-based user guide. Others have produced an Adobe Acrobat PDF document from their paper guide. In both cases, the content of the user guide and the Help have been identical.

The benefits of Level 1 single sourcing include:

- Reduced creation and production costs
- Reduced development time

This type of single sourcing, however, has drawbacks in the area of usability, as information used in one type of documentation is not necessarily well suited for use in another. It also raises the issue of the suitability of the materials for an online environment. This is particularly true for the PDF file, as it is essentially a paper-based document provided online. The PDF files are rarely optimized for use online, rather users tend to print the guide and use a paper-based copy.

Level 2: Static customized content

Level 2 single sourcing provides much more opportunity for effectively designing information. This type of single source material is customized to meet the needs of the user, the type of materials to be developed, and the output media. Note that this customization does not involve altering the source elements. Rather, specific elements are included or excluded depending upon the requirements and the media used, and additional content is added where necessary. Examples of Level 2 single sourcing include the following. In each case, there is a core of content in common (the single source), as well as customized content.

- **Multiple media output**
  The content is customized to meet the specific media capabilities (e.g., Help vs. paper vs. Web).
- **Multiple platforms**
  The content is customized to reflect the different content based on the platform the information is intended for (e.g., NT vs. UNIX)
- **Product families**
  Frequently corporations sell product families, or a suite of products. The content is customized to reflect the product-specific components of the product suite.
- **Multiple information products**
  Corporations rarely create just one type of information.

More often, they produce multiple information types, such as User guide, Help, Reference card, and Training. These are known as information products. The content is customized for each of these information products.

- **Multiple audiences**
Information sets can have multiple audiences. Some information is relevant to all audiences, and some information is specific to a particular audience. The content is customized for each audience.
- **Multiple releases**
Products go through release cycles. Documentation must reflect these releases. Some corporations maintain parallel releases, others have consecutive releases. In each case information builds on the existing information set, the source.
- **Integrated documentation and training**
User documentation is task based. Training materials are also task based, though they add information elements like examples, exercises, and objectives. Training materials build on the core content of user documentation, the tasks.

The benefits of Level 2 single sourcing include:

- Increased usability of information for users
- Reduced development and maintenance costs
- Reduced costs of translation (content is only translated once)

Level 3: Dynamic customized content

Dynamic customized content is information that is assembled only when it is requested. It does not exist as a document; rather as a series of information objects that are assembled in response to the user's requests or requirements. This is similar to the automatic assembly of relevant information for e-Commerce applications; however, dynamic content is more comprehensive, usually involving entirely customized documents.

Note that although the majority of dynamic content is delivered dynamically through the web, you can also dynamically assemble a document, and then provide it as a PDF when paper output is required. In addition, web-based learning materials can be created dynamically, providing a unique learning experience for each learner.

Dynamic content provides the following benefits to corporations:

- Ability to create multiple information products on demand
- Specifically address customer needs
- Reduces the cost of creation of multiple information products

Dynamic content provides the following benefits to customers:

- Reduces or eliminates the need to search for relevant information
- Provides customized content
- Provides "just-in-time" content
- Provides content which reflects their requirements or system configuration
- Ability to customize the content further to reflect their nomenclature and unique processes

The role of single source levels in information modeling

Each increasing level of single sourcing brings with it increasing levels of complexity in the structure of the information, and therefore the information models, as well.

Level 1 single sourcing requires simple models, but Level 2 static customized content requires that you match models from one information product to another to determine reusable elements across documents. A Level 3 dynamic content engine requires detailed information product models so that the engine can programmatically interpret the customer's needs, and assemble a document that is effective and usable.

What is information modeling?

One of the most critical phases of single sourcing is building the information models on which the project is based. During the information modeling phase, you determine the elements required for each information product (or output), and how each information product will be designed for optimum usability and reuse. Once you decide which information products you need to develop and the information they should contain, you can develop the metadata, content standards, and templates to support them. The information models become, in effect, the road map for your project and provide the framework that authors fill in.

In single sourcing, you break information down to the element level (e.g., section, paragraph, sentence). Instead of writing documents, you write elements that are stored in a single source (often a database). Elements are then compiled into information products from that single source. The power of single sourcing lies in effectively reusing information elements—whether they are paragraphs, procedures, or sentences—over and over again. Information models identify all the required elements and illustrate how to structure and reuse them.

The process of information modeling involves identifying all the information requirements for a particular project or an entire organization up front, then building a model that illustrates how information elements will be compiled to form each information product. Authors refer to models to determine which elements are required for each information product, and which elements are reused.

Designing effective models

Designing effective models requires that you start with a solid audience and information analysis to understand who uses what information, and in what context. Information models should depict all possible uses and users of your information, so you need to understand their needs up front. Thorough analysis of both use and users forms the basis for your information model.

Audience analysis

During audience analysis, you create detailed audience profiles for those who will be using the materials:

- Conduct a detailed analysis of the tasks the users will perform (what are they expected to know or be able to do?)
- Get clear descriptions of customer goals and expectations for the project; consider both internal and external customers

Information analysis

During information analysis, you look for:

- Repetitious information
- Similar information
- Potential missing information
- Multiple outputs (how many different formats does the information currently appear in? paper? online? brochure?)
- Multiple formats (table? bulleted list? paragraph?)
- Multiple audience requirements
- Information product commonality

Types of models

There are two types of models you need to create:

- Information product models
- Element models

Information product models

Information product models identify how information will be compiled to create an entire information product. They contain information such as which elements are required, which are optional, and the order they appear. For example, you could have an information product model for:

- Reference materials
- User guide
- Training

- Application notes
- Installation guide

The top level of an information product model of a software installation guide might include:

- Cover
- System requirements
- Installing Product ABC
- Uninstalling Product ABC
- Index

- Front matter
- Installation Checklist
- Troubleshooting
- Glossary

Element models

Element models identify how each element in the information product model is structured. They identify such things as the semantic structure for each element, the metadata associated with each element, as well as any other elements nested within elements.

For example, the Installing Product ABC section would have the following elements:

Installing Product ABC
Introduction
    Procedure

The procedure element could be broken down further:

Procedure title
    Link to previous procedure
    Prerequisites
    Procedure body
        Introduction
        (Notes, cautions, warnings)
        Step
                Step explanation
        Link to next procedure

The procedure element is known as a *container element*. Procedure body is also a container element, as is Step. Any element that contains other elements is referred to as a container element.

This procedure model would probably work for a user guide or maybe Help, but if you also needed to create training, this procedure would not suffice. The model could be modified to accommodate all the outputs in the following way (table 12.1).

In this case, the Help is designed to be quick reference, the user guide provides more explanation, and the training provides objectives and exercises.

A comprehensive procedure model could be provided in as shown in table 12.2 (next page).

Nested models

The procedure element is known as a nested element. Help is nested within the User Guide that is then nested within the Training. Figure 12.1 (next page) illustrates how the nested model could be converted into a variety of outputs and media.

Table 12.1. Accommodating multiple outputs with a model.

Help	User documentation	Training
Procedure title	Procedure title	Procedure title
X-ref to previous procedure	X-ref to previous procedure	
	Prerequisites	Prerequisites
	Introduction	Introduction
(Notes, cautions, warnings)	(Notes, cautions, warnings)	
		Objective
Step	Step	Step
	Step explanation	Step Explanation
Field description	Field Description	Field description
		Exercise
X-ref to next procedure	X-ref to next procedure	

Table 12.2. Identifying appropriate outputs for elements

Element	Valid for
Procedure title	Help, User Guide, Training
X-ref to previous procedure	Help, User guide
Prerequisites	User guide, Training
Introduction	User guide, Training
(Notes, cautions, warnings)	Help, User Guide
Field description	Help, User Guide, Training
Objective	Training
Step	Help, User Guide, Training
Step explanation	User Guide, Training
Exercise	Training
X-ref to next procedure	Help, User Guide

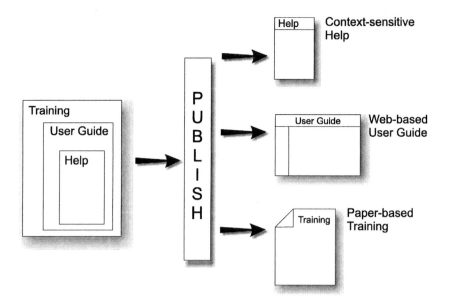

Figure 12.1. Nested content.

A nested model makes it possible for authors to clearly see how reusable elements fit together. For example, if your organization has documentation and training in separate departments, a nested model makes it easy for the original content (source) to be written by the documentation group, then have the training group easily reuse the content in their training materials by simply adding objectives and exercises to the existing tasks. Alternatively, if your organization has a combined training and documentation group, you could have a single author create all the content (both training and documentation) for a specific task.

Building block models

Nested models are appropriate when content across information products is closely related. However, you can also single source content from very diverse information products. In this case there will be some elements in common, but not all. For example, table 12.3 (next page) illustrates some common elements across a number of information products.

When creating content using building blocks, the author creates the unique elements of content, and searches the content management database for existing content identified as a reusable element for reuse in their content.

Understanding granularity

Single source information consists of elements, not files or documents. You need to identify the elements of your information set. To do this, you look at the *granularity* at which information can be broken down. Granularity means looking at the smallest possible piece of information that is still a useful piece of content.

You need to scrutinize your models to ensure the level of granularity is appropriate to your customer needs, your author requirements, and the level of maintenance required to maintain your information set.

The issue of how small to break down the model is the issue of granularity. Granularity involves identifying the smallest possible piece of information that is still a useful piece of information. A piece of information is useful if you want it to:

- Ensure consistent content by authors
- Identify the element for retrieval
- Identify the element for reuse

Identifying this content so that authors can consistently create it is valid; however, you are unlikely to want to retrieve or reuse these elements. Typically, elements that contain content that is smaller than a paragraph

Element	Press release	Data sheet	Brochure	Help	User Guide	Training	Web Site	FAQ
Product description	X	X	X		X	X	X	
Benefits	X	X	X				X	
Contact information	X	X	X				X	
Field definitions				X	X	X		
Trouble-shooting					X			X

Table 12.3. Building blocks.

such as sentences and words are not recommended. Very small elements make it difficult for authors to create and administrators to maintain.

There are exceptions to this. For example, you may need an element that identifies a product name, or a field name. The product name may not be determined when you start to write so you want to have an element that can be changed very easily. Or you may have software that can be customized where the customer may change the name of a field, so you need to be able to identify that element. These types of elements are often referred to as variables or entities. An entity typically has a definition (meaning) that can be changed in one location and automatically updated throughout your content set.

For example, elements like Warning could be broken down further.

Warning
• Signal word/icon
• Title
• Warning body
    o Hazard avoidance
    o Hazard identification
    o Consequences

If you were to write a warning to this model, it might look like this:

Warning
Throw your PFD away if it is faded.
Do not continue to use a PFD that is faded.
A weathered PFD could tear easily, resulting in loss of flota-tion material.

However, a better model might be:

Warning
• Signal word/icon
• Title
• Warning body

It would probably be better written like:

Warning
Throw your PFD away if it is faded. Do not continue to use a PFD that is faded. A weathered PFD could tear easily, result-ing in loss of flotation material.

You would probably not break the Warning body down as was done in the first model, because the identified elements would be contained within one element, a paragraph. Include a description of these elements in your model to provide guidance to the author.

Semantic elements vs. base elements

The word semantic, means *meaning*. A semantic element is a specific element that uniquely identifies the content of that element. In other words, the name of the element uniquely identifies the element. Semantic elements make it easy for the author to identify exactly what content they should include. Semantic elements also act as metadata and enable you to identify and reuse specific content.

A base element is an element that specifies a common name of an element, such as paragraph (para) or unordered list. Base element names draw on SGML standards. Base elements are also used to provide guidance to the stylesheet/template creator to indicate the structure the format can contain. Common base elements include:

Element	Description
title	title
para	paragraph
ordered list	ordered list (numbered list)
itemized list	unordered list (bulleted list)
table	table
figure	figure/graphic/illustration
note	note
tip	tip
warning	warning
caution	caution
xref	cross reference
link	hypertext link
ulink	link to a URL
glosslist	list of glossary terms and their definitions

For example, assume, you need to retrieve an Introduction for Product ABC. If content is semantically tagged the search engine will search only Introductions that either have Product ABC mentioned in the text or that have the Product ABC metadata attached to it. However, if the majority of text is tagged with a base element tag like para the search engine must search the entire information set and will probably return inappropriate results. The author must then spend additional time manually searching through the results until they find the correct "hit."

Additionally, if you asked an author to create a procedure and you gave them a model that looked like the Base element model shown in table 12.4, they would find it very difficult to correctly create a procedure.

Table 12.4. Base element model vs. semantic model

Base element model	Semantic model
title	Procedure title
para	Xref to previous procedure
para	Prerequisites
para	Introduction
para	(Notes, cautions, warnings)
para	Field Description
Itemized list	Objective
Ordered list	Step
para	Step explanation
para	Exercise
para	Xref to next procedure

However, not every element can be semantic. The more structured the content and more reference oriented the easier it is to define semantic elements. The less structured the content the harder it is to define semantic elements. A procedure is an example of structured content, but an introduction is not. Your introduction may look like:

    Introduction
        Para
        Itemized list

Where possible you should use semantic elements. However, if you are not using a structured editor, semantic elements should be used carefully as too many semantic elements can result in too large a list of tags for authors to choose from. (See the section on *Formalizing your model.*)

Frequency of element usage

If you are creating XML models or XML-ready models you need to define the frequency of each element. If you look at the Procedure element, you can see that Step is listed only once. However, it is very unlikely that your procedure will have only one step in it. Therefore you have to indicate how frequently the Step can occur.

Table 12.5. Frequency notation.

Notation	Description
	When no notation is used with an element, this implies that it is mandatory
+	Repeat one or more times
?	Optional (0 or 1)
*	Optional (0, 1, or more)
|	Or
( )	Optional, any item in any order. This is usually expressed by putting the items in an oval

SGML (Standard Generalized Markup Language), the precursor to XML has defined the standard notation for frequency of use (table 12.5). The frequency symbol is added to the element name.

The Procedure model would have the following frequency:

- Procedure Title
- Xref*
- Prerequisites
- Introduction
- (Notes, cautions, warnings)
- Field Description*
- Objective+
- Step+
- Step explanation
- Exercise
- Xref*

As you can see, for the most part, each element can only occur once (no frequency symbol). Elements like Objective and Step can repeat one or more times. The cross references (Xref) and Field Description are optional. Notes, Cautions, and Warnings are optional, but when they are used they can appear in any order.

Information product usage

The frequency of usage indicates the frequency of usage of an element. You also need to define the usage of elements in information products.

Some elements are identified as optional. In some cases an element may be optional in one information product and mandatory in another. For example, the Xref was marked as optional in table 12.5. However, you may choose that the cross references are mandatory in the Help, optional in the User Guide, and not required at all in the training. This type of usage needs to be marked differently in your model. This is usually indicated by an O (Optional) or M (Mandatory).

A mandatory element is only mandatory if the container element above it is used. For example, if an element is optional, and the author decides to use it, any mandatory elements below it must also be included. However, if the writer decides *not* include an optional element, then the elements below it are not used, even if marked mandatory.

The information product usage of the procedure model would be illustrated as in table 12.6.

Tree diagrams

A tree diagram is a visual representation of your model. The tree diagram is also based on SGML modeling. It is not a mandatory component of the modeling exercise but it is useful. Many authors understand models better by viewing a visual representation of the model and

Table 12.6. Information product usage.

Model	Help	User Guide	Training
Procedure Title	M	M	M
Xref*	M	O	
Prerequisites		M	O
Introduction		M	M
(Notes, cautions, warnings)	O	O	
Field Description*	M	O	O
Objective+			M
Step+	M	M	M
Step explanation		O	M
Exercise			M
Xref*	M	O	

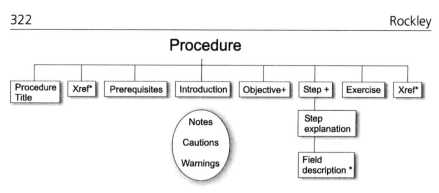

Figure 12.2. Sample tree diagram.

creating a visual version during modeling may assist your team in
identifying problem areas. Figure 12.2 is an example of a tree diagram.

Metadata

Single source content needs some way of classifying and identifying all
of the information or content "bits" so that they can be retrieved and
combined in meaningful ways for customers.

What is missing is information about the information—that is, the
labeling, cataloging and descriptive information that allows the content
elements to be properly processed and searched by a computer. This
information about information is known as *metadata*.

What is metadata?

Traditionally, metadata has been defined as "data about data". Metadata
enables content to be retrieved, tracked, and assembled automatically.
Metadata makes content accessible. Metadata enables:

- Effective retrieval
- Automatic routing based on workflow status
- Tracking of status
- Reporting

Main uses of metadata

There are generally three activities you perform in relation to the content:

- Reuse
- Retrieve
- Track

Metadata for reuse

Metadata for reuse can be particularly useful in a single source strategy, eliminating content authoring redundancies. In this case, metadata is applied to each content element. Authors can search for elements before beginning to write, to see if they already exist somewhere in another document stored in your content management system.

Metadata for reuse could include:

- Content type
- Where the content should appear
- Creation date
- Content owner
- Keywords
- Links to where content is already used

Metadata for retrieval

Metadata for retrieval enables content to be retrieved through searching, either in an authoring tool, or in your retrieval tool, such as your intranet or the Internet. Metadata for retrieval can includes much of the same metadata you define for reuse, but is usually much more extensive. It can include metadata such as:

- Title
- Author
- Date (creation, completion, modification)
- Keywords
- Responsible party
- Security status
- Tracking (e.g., status)

Metadata for retrieval enables users to specifically define which content elements they want to view. This metadata can also be used to dynamically populate content for users, based on specific profiling information.

Metadata for tracking (status)

Metadata for tracking is particularly useful when you are implementing workflow as part of your single sourcing solution. By assigning status metadata to each content element, you can determine which elements are active, control what can to be done to an element, and who can do it. Generally, status changes based on the metadata are controlled through workflow automation, not by end users.

Status metadata can include:

- Draft (under development by the author)
- Draft for review
- Reviewed
- Approved
- Final
- Submitted

Defining metadata and metadata categories

Properly defining and categorizing the types of metadata you want to capture about your information is extremely important to the success of your single sourcing strategy. First of all, you need to determine if you are defining metadata for retrieval, reuse, or tracking. Then, you need to understand what the end business result is that you are trying to achieve, and build your metadata backwards in order to achieve that result. For example, if you want to have a web site that dynamically displays the correct instructions to a user based on their level of expertise, you will need to identify all possible information, or "metadata," that could be involved in determining the specific content the user should view. For example, you will need to identify:

- All user levels (e.g, novice, intermediate, expert)
- All types of users (e.g., customers vs. systems administrators)
- System the user is using (web site to access information vs. system administration)
- All content types viewable by each user type
- Content type classifications (e.g., subject, level of expertise)
- User location

Then, you need to determine which business rules apply, for example:

- What content type each user type is allowed to see
- Which system's content each user type is allowed to see

Other considerations

Care should be taken not to design too much metadata, which can produce too many search results, or too little metadata, which can result in inefficient retrieval. Properly defining the metadata you need helps to make sure that the right information is delivered to the right person, for the right reason, at the right time.

Sample metadata

In the example of the procedure, the most obvious metadata is that for information product type:

- Help
- User Guide
- Training

For more examples of metadata refer to Table 12.7 (next page).

Recognizing patterns in modeling

As you model, you will find that there are elements that are used again and again in different information products. For example:

- Introduction
- Procedure
- Warning
- Glossary

These are known as Common Elements. You can define these elements once and refer to them from different information product models. After you have created a number of element models, you may notice that they have the same structure. In fact, they could be identical except for their name. For example:

- Warning
- Caution
- Note

All have the following structure:

- Signal word/icon
- Title
- Body

In this case you could create a single element such as Alert that has the following structure:

Alert
    Signal word/icon
    Title
    Alert body

Table 12.7. Sample Metadata.

Attribute	Value	Default	Selection	Applies to
Status	Draft	Draft	Single	All
	Draft for review			
	Reviewed			
	Approved			
	Final			
Image type	Screen	None	Single	Images
	Logo			
	Flowchart			
Alert	Warning	Note	Single	Alert
	Caution			
	Note			
	Important			
	Tip			
	Warning			
InfoProduct	All types	All types	Multiple	All
	Press release			
	Data sheet			
	Brochure			
	Help			
	User guide			
	Training			
	Web site			
	FAQ			

Table 12.7 (cont'd). Sample Metadata.

Attribute	Value	Default	Selection	Applies to
Audience	Internal	All	Multiple	All
	System administrator			
	Developer			
	Web master			
	End user			
Step type	Procedure	Procedure	Single	Step
	Exercise			
	Review			

And use metadata to define what type of alert it is:

Alert type
    Warning
    Caution
    Note

Formalizing your models (style guide)

You need to formalize the structure of your models. The formalized structure is used to create your templates, develop your writing guidelines, and potentially map your information to XML. Formalized models are typically created using a spreadsheet program and a graphing program (table 12.8, next page). A formalized model consists of:

- Tree diagram (optional)
- Semantic model
- Style mapping (where appropriate)
- Base element (using standard element structure and tree diagram)
- Information product usage
- Metadata
- Writing guidelines
- Production guidelines
- Example

Table 12.8. Sample formalized model.

Semantic model	Style mapping[1]	Base element	Help	User Guide	Training	Writing guidelines
Procedure Title	Procedure title[2]	title	M	M	M	Title for the procedure. The title should be task-oriented.
Xref*	Normal	link	M	O		Link/cross-reference to the logical preceding procedure.
Prerequisites	Normal	para, itemized list		M	O	Description/list of prerequisites required before performing this task.
Introduction	Normal	para, itemized list		M	M	Introduce the procedure. Indicate when the user would perform this task.
(Notes, cautions, warnings)	Common model[3]	Common model	O	O		See common models.
Field Description*	Common model	Common model	M	O	O	See common models.

[1] This column is only required if your authors are not using a structured editor.
[2] The semantic Procedure Title is retained to assist in some retrievability.
[3] These elements are described elsewhere as they are common to many other container elements.

Table 12.8 (cont'd). Sample formalized model.

Semantic model	Style mapping1	Base element	Help	User Guide	Training	Writing guidelines
Objective+	Normal (bulleted)	para, itemized list			M	Objective for the task. Introduce the objective, then follow the introduction with an unordered list of specific objectives.
Step+	Normal (numbered)	ordered list	M	M	M	Steps for the procedure. The step should be the action only. Explanation of the step should be included in the explanation.
Step explanation	Normal	para		O	M	Explain the step in further detail. This is optional and only required when further explanation is required.
Exercise	Normal	para, itemized list			M	Relevant exercise for the user to perform.
Xref*	Normal	link	M	O		Link/cross-reference to the logical following procedure.

At this point, I have covered all the items listed earlier except Style Mapping, Writing Guidelines and Example. These sections are only applicable to a style guide, not part of the process.

Style mapping

As indicated in *Semantic models vs. Base element models,* semantic tags can cause problems if you are not using a structured editor. A structured editor will automatically present the author with the next appropriate element to use, and will restrict the number of elements that can be selected based on the container element selected. However, in a standard word processing tool like Microsoft Word or desktop publishing tool like FrameMaker (unless you are using Frame+SGML), this is not the case. The result is that you have to use a large number of semantic tags, which results in large volumes of style tags. It can be an onerous task for authors to select the appropriate tag from a large list. Therefore, you should still have a semantic model, which enables your authors to clearly understand what they need to put into the content, but indicate the non-semantic style tag they should use in their editor.

Writing guidelines

No matter how clear the model, you should always include writing guidelines to assist the author in understanding what they should create. Provide any tips or rules about how to write the element.

Example
Always provide a sample of completed content using the model. The sample should be a best practices example that can help to illustrate to the author how the desired content should be written.

Production guidelines

Production guidelines provide recommendations about how the element should be displayed in the output. This information should only be in the materials provided to the person responsible for templates or stylesheets, not the author.

XML and single sourcing

Although XML is not necessary for single sourcing, it does facilitate single sourcing. For complex, Level 3 single sourcing (dynamic content), XML is the technology of choice. XML supports the chunking of information into elements down to the paragraph or even sentence level. This chunking, along with efficient use of metadata, enables more efficient search and retrieval of content elements.

What is XML?

XML is a standard for the development of markup languages for web-based information. If you are familiar with HTML, you are familiar with an instance of markup languages. But unlike HTML, which has a fixed set of tags, XML lets you define your own markup language.

Based on SGML, XML has been optimized for web-based delivery of all kinds of information. A family of standards, rather than a single standard, XML defines all aspects of information presentation, including markup, linking, style, structure, and metadata. XML is fast becoming the new Internet standard for information exchange.

Benefits

Using XML has a number of benefits that support single sourcing:

- Reuse of existing content, reducing redundancy and costs
- Dynamic content delivery of personalized content
- Separation of content from presentation to allow single sourcing of multiple output formats
- Better-managed content, resulting in reduced costs
- Improved search and retrieval capabilities through metadata use.

XML and structured content

XML requires a DTD (Document Type Definition) to support the development and management of content. The DTD is like a structural template: it explicitly defines the structure of the content. This explicit structure ensures that authors can only enter content which follows the structure. That is, all of the required pieces of information are in place and in the correct order. There is no possibility of entering invalid information. This will assist authors in writing rapidly and eliminate validation errors.

Separation of content and format

Authors using products like MS Word are used to applying style tags that define the "look and feel" of the content. XML tags define the content. This means that any look and feel can be applied to the content, depending upon the desired output. For example, the content can look one way on paper, another in HTML, and many other ways if used in an article, presentation, or poster.

The look and feel is defined by the appropriate stylesheet selected in the final production process, or at any time in the authoring/review cycle.

Built-in metadata

In XML, you decide the tag names that you'll use in your documents. So you can create semantic tags (tags that have meaning), rather than generic tags. You apply tags that describe the content of the information, not the formatting. For example, typically in MS Word there is a tag labeled "Normal" that you would apply to information that you want to be formatted in a certain way. In XML, tags could be called "introduction," "title" or "objective". The semantic tags automatically provide metadata about the content they enclose, and can be interpreted for display in many ways.

For explicit metadata, XML can define attributes for the elements in a document. Similar in syntax to HTML attributes (color='red'), XML attributes are defined by you, to provide whatever additional information is required to identify the use of the information. Attributes can by used to identify information that is specific to format, product, user, or use.

Use of XSL

XSL (eXtensible Stylesheet Language) is the piece of the XML family of standards that defines formatting. But unlike a traditional stylesheet, which manages the look of a document, XSL is used to convert XML documents to other formats. These include HTML for web output, other markup languages like Wireless Markup Language (WML), and PDF. In addition, XSL stylesheets can be used to manipulate information, including sorting, filtering, moving and repeating information, and generating new information, such as tables of content and lists of figures.

For single sourcing, you can create a stylesheet for each required output product and pass the same XML file through each, automatically generating output that can differ in format and content.

Virtual documents

XML enables you to build documents out of individual content files on the fly. Individual pieces can be assembled upon demand, in response to user requests or to meet the needs of a specific output format.

Using an information model

Once you have your models in hand, there are a number of ways in which the models can be used.

Information technologists use the models as a specification to create the templates and DTD where appropriate. In addition, if you are using a content management system, the models are used as the specification to set up the system that will control the content.

Authors use information models to determine what information goes in which information product, as well as how to structure each element. Referring to the information model, they can determine, for example, that an information product requires error codes, the structure of an error code, and get hints/rules (writing notes) about how to write an error code. The information model ensures that the reusable content is consistently structured for transparent reuse.

Model reviewers check information models to ensure that the models will effectively support customer and information requirements.

Reviewers use information models to review authors' drafts. They compare the draft against the information model to ensure that it contains all the necessary elements. They also review the models to ensure that they contain all the necessary elements for each information product.

Tools use the information models to control content. A structured authoring tool (e.g., XML editor) uses information models as the template (e.g., DTD) to guide the author in creating content. Using the model, the system prompts for the next valid element. Dynamic content engines use information models to identify which elements can be validly combined and in what order, to create a usable information product for users when they request it.

Summary

Information models are critical to successful single sourcing. Effective information models serve to identify all the knowledge within an organization, and to capture and reuse it effectively.

The information modeling process forces you to consider all information requirements and to assess what information is available to fulfill those requirements. The model also ensures that authors develop information in the same way, so it can be reused effectively.

The information model becomes the "catalog" of all information products produced within an organization, and outlines the necessary information elements for each of them.

Information modeling checklist

Identify all possible uses for the information. Ask questions like:
- ☐ Who will use the information? Do their needs differ?
- ☐ Where is the information required (e.g., internet? paper?).

Determine the level of single sourcing required. The 3 levels are:
- ☐ 1 – Identical content, multiple media
- ☐ 2 – Static customized content (author customizes)
- ☐ 3 – Dynamic customized content (user customizes)

Create detailed audience profiles:
- ☐ What tasks do they perform?
- ☐ What are they expected to know/do?
- ☐ What are the customer's objectives for the users?

Conduct detailed information analysis. Look for:
- ☐ Repetitious information
- ☐ Similar information
- ☐ Missing information

Identify the granularity of the information. Ask:
- ☐ What is the smallest piece of information that will be useful?
- ☐ What information is core?
- ☐ What information will change?
- ☐ At what level will it change?

Build models:
- ☐ Create a model for each information product.
- ☐ Create a model for each information element.

Identify structure.
- ☐ Identify the hierarchy of structure so you know how the elements fit together.
- ☐ Create tags so you can identify elements according to content.

Formalize structure. Create:
- ☐ Semantic model
- ☐ Base element model
- ☐ Writing guidelines for each element
- ☐ Style mappings so authors know what styles to use for each element
- ☐ Metadata to further describe content
- ☐ Examples of how each element should be written

Suggested Readings

Kostur, P., & Rockley, A. (2001, May). Information modeling for single sourcing, *Proceedings of the 48th Annual STC Conference.*

Maler, E., & El Andaloussi, J. (1996). *Developing SGML DTDs: From text to model to markup.* Upper Saddle River, NJ: Prentice-Hall.

Marco, M. (2000) *Building and managing the meta data repository: A full lifecycle guide.* New York: Wiley.

Rockley, A. (2001) *Enterprise content management through single sourcing.* Whitepaper [On-Line] Retrieved July 2001 from the World Wide Web: www.rockley.com.

Rockley, A. (2001). Dynamic content management. *Intercom*, February.

Rockley, A. (2001). The role of single sourcing and technology. *Technical Communication, 48*(2), 189-193.

Rockley, A., et al. (Fall 2002) *Managing enterprise content: A unified content strategy.* Indianapolis, Indiana: New Riders Publishing.

13

Redesigning to Make Better Use of Screen Real Estate

Geoff Hart
Forest Engineering Research Institute of Canada

Although complexity is often inherent to certain information design problems, and that complexity relates strongly to the information content in others, sometimes what seems to be complexity is nothing more than a poor choice in representing the information content. These poor choices may take several forms, including:

- Poor grouping and chunking of the content, so that related information is widely separated and unrelated information is grouped together (Bertin, 1983). By conforming with *gestalt* principles (e.g., Moore & Fitz, 1993), good grouping supports the viewer's ability to recognize like and unlike objects.
- Redundant repetition of labels and other affordances. In this chapter, I've chosen to define *useful* redundancy as any repetition that supports performance of the user's tasks rather than interfering with this performance. Although redundancy and repetition are often useful techniques in instructional design because they assist the process of moving information from short-term memory into long-term memory (e.g., Coe, 1996), they are less helpful where space is tight and the goal is instead to provide aids to navigation and comprehension.
- Poor sequencing of the information with respect to the manner in which the user will access that information. Correcting

337

that structure to follow a more logical design (e.g., Mandersloot, 1996) can greatly improve the effectiveness of communication, particularly when the corrected structure more closely follows the sequence followed by users of the information or helps those users follow an efficient path through the information rather than having to develop that path themselves (Horton, 1994).

Software screen design provides a common example of the kinds of redesign challenges that can arise from failing to recognize these problems. Although it is true that electrons are cheaper than paper, and you can potentially use any number of sequential screens to guide users through a task, human patience is a limited resource. Thus, it is often wiser to use as few screens as possible to get the job done. Moreover, sometimes you really do need to stick to a single screen because of limited developer resources or other factors. This doesn't mean that you can never use more than one screen to obtain user input, or that you should cram information densely into a screen simply to reduce the number of screens, but it does mean that you should look long and hard at any given interface design to see whether you can simplify it to make better use of the available space.

This is particularly relevant given that many technical communicators are beginning the transition into producing usable designs for handheld computers such as personal digital assistants (PDAs) or cell phones, where screen space is extremely limited. Software developers generally lack the design skills to fit information useably in a constrained space, and often turn to us in the hope that our expertise with words will help. Often, their first request is to simply edit the names of data-entry fields and other labels so they can fit within a specific number of characters; in practice, the actual goal more commonly turns out to involve making better sense of a proposed design and thereby reducing the need for purely mechanical tricks such as shortening the text. Other times, we're called upon to document an ineffective interface that overwhelms users with poorly designed labels, a dense screen of text, no clear sequence, and missing affordances (clues to the meaning and use of fields, buttons, and other objects on the screen). In either case, we face an interesting challenge in information design.

Fortunately, we have access to simple yet powerful techniques for analyzing and solving such problems, often without requiring major reprogramming of the interface. That latter point is important, as programmers are often reluctant to make significant interface changes solely on the advice of a writer or editor. With modern screen-design software, the task of reordering and relabeling screen elements becomes substantially easier, enough so that programmers are more willing to work with us to redesign screens.

The analysis process I propose for improving screen designs involves four steps:

1. Identify and group fields that have considerable information in common: Using the information common to each field as a heading for *all* the fields eliminates the need to repeat the common text beside each field, thereby reducing the amount of text to be read. This heading also provides a single context in which the grouped information can be absorbed, thus potentially reducing the cognitive load (e.g., Cooper, 1998) on the viewer.
2. Examine what remains after extracting the common information to identify possible affordances and alternative presentations of the information: With the heading providing the necessary context, the labels can often refer more directly to the concept they introduce or represent, and this often shortens them further. Moreover, even though text is sometimes necessary, graphical presentations of information, such as checkboxes and radio buttons, may accomplish a goal more effectively.
3. Identify the typical or likely sequence of activities and seek dependencies: In some designs, the choice entered in one field constrains the permitted choices in other fields; in other designs, selecting one choice predetermines the information to be used in subsequent fields, thereby eliminating the need for users to fill in those fields. In both cases, matching the sequence on screen to the sequence of the user's actions should provide benefits in terms of improved usability because users spend less mental effort adjusting their approach to match that imposed by the software.
4. Repeat these steps iteratively to determine whether further simplification is possible.

In this chapter, I have illustrated how these principles might work by means of two examples: one hypothetical but based on a real situation, the other one based on an actual product. Each step reiterates the fundamental components of the analysis that underlies good information design: thinking about a problem from the audience's perspective; thinking about the underlying principles of how something works; simplifying a design based on the relationship between the audience's needs and how something works; and confirming that the results really do suit the audience's needs.

Example 1: Reducing text length

Group similar fields and extract commonalities

A question that appeared in the techwr-l technical writing discussion group (join this group via "Subscriber central" at www.raycomm.com) involved simplifying the labels for four fields related to a single function in the user interface. All four fields involved variations on the same theme, and thus needed to appear both similar, to reflect the common theme, and distinct, to reflect the slight differences in meaning. The specific context was a database of contracts, each with an expiry date, and the user's need to find specific contracts in the database. Furthermore, the developers insisted that the labels be no more than three words long, a particularly tight space constraint.

In this case, the similar fields had already been grouped by the developers because all involved the need to find a contract based on its expiry date. The remainder of this first step in the analysis was to identify the common components that could be removed and used as a heading. The four fields resembled the following:

- Contracts that will expire within 30 days from today
- Contracts that expired less than 14 days ago
- Contracts that will expire more than 60 days from today
- Contracts that expired more than 30 days ago.

The information that was common to each field fell into two categories based on the fact that users were searching for contracts, and based on the expiry dates of these contracts. Of these, some contracts expire in the future, whereas some have already expired. The writing challenge became that of choosing a verb form that worked well for both situations. To avoid repeating the common words in each field, the redesign can begin by creating a heading that combines all three concepts. A good first effort would be "Display contracts with expiry dates:", a heading that implies through the presence of a terminal colon that a list of options for defining the dates follows. This heading now applies equally well to all the fields, and need no longer be repeated for each field. After extracting the common information to produce a heading, what remains is the portion of the interface that defines the expiry date (figure 13.1).

Seek affordances

The second step involves re-examining the design constraints to discern the actual problem we're trying to solve. Here, the problem lies in the

Figure 13.1. An initial simplification that results from extracting the common information.

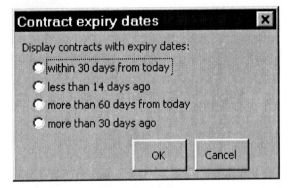

assumption that "a maximum of three words" is literally correct. It isn't. In reality, the developers have simply chosen this terminology to suggest that *space* is limited, not that the words themselves are limited. For example, it is doubtful they would accept a solution that uses a single 60-letter word or two 40-letter words, even though these meet the strict definition of the problem that they proposed; conversely, the screen design might well leave plenty of room for five three-letter words. This being the case, the design constraint should be rephrased so as to express the actual space permitted. Because this chapter deals with a hypothetical analysis, it wasn't possible to define the actual space constraints, and for the sake of my example, I've modified the original request. Assuming that English root words average about five letters long, the "three words" constraint suggests a design goal of no more than 15 letters per field. (In a real example, we would of course confirm this limit with the developers.)

The redesign isn't complete because all four labels in figure 13.1 are still longer than 15 characters. (The original wording of the design question is also unclear, because the actual meaning may be "less than or equal to" and "greater than or equal to." To simplify my example, I've assumed that "less than" and "greater than" are literally correct.) Judicious use of symbols such as < for "less than" and > for "greater than" might help, but we would first have to confirm that our audience understands these symbols. Many members of a general audience may never have seen the symbols before, or may not have seen them since their last math course in high school, potentially decades ago. That being the case, relying on these symbols could require the use of popup explanations such as the "tool tips" that programmers use to identify the roles of icons, or finding another efficient form of context-sensitive help that ensures users can learn the meanings quickly if they don't already know them. Because that is one step removed from understanding the design "at a glance," and requires a more complex implementation that the developers may be unwilling to program, it's preferable to use words rather than symbols.

Seeking affordances can help resolve the problem. In this phase of the analysis, we should consider whether text labels are truly the most effective approach and whether other affordances might work better. For example, we could redesign the previous list using checkboxes based on recognizing that the simplified labels used in the first revision of the fields contain two additional commonalities that could become subheadings: somewhat verbosely, these would be "future items versus past items" and "more than versus less than." With a little modification, these provide suitable subheadings that further improve the design (figure 13.2).

Identify opportunities that arise from dependencies

In the third phase of the analysis, we try to identify the most typical sequence of activities and any dependencies that arise from this sequence. For example, users might typically determine the envelope within which the problem occurs (how many days) and whether the situation lies in the future or the past. This would lead to an improved design that takes advantage of two additional principles: The number of days can be displayed only once, and the direction (before or after today) can become a simple choice, as shown in figure 13.3. The advantage of this new design is that it expresses the results of the user's choices in a form that mimics how users themselves would express the problem. Moreover,

Figure 13.2. An improvement based on modifying the design in figure 13.1 to take advantage of affordances.

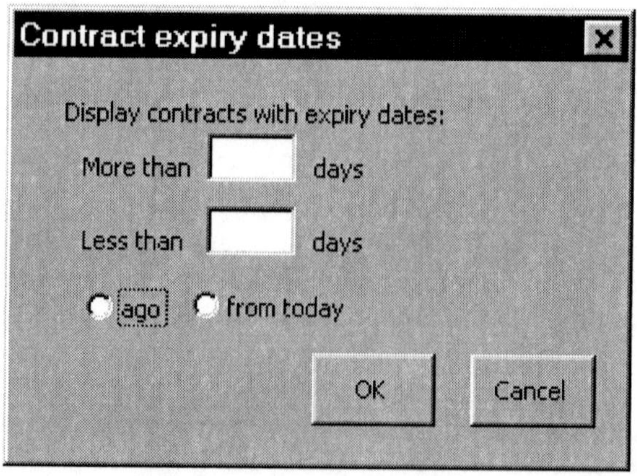

Figure 13.3. An improvement on figure 13.2 using affordances and a recognition of the likely sequence of activities to complete the task.

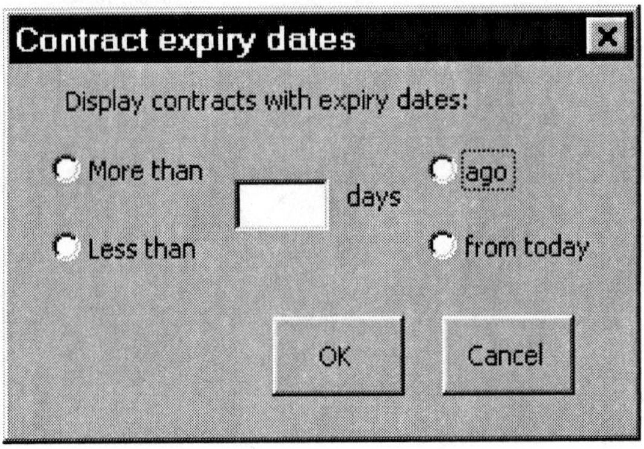

the screen layout presents the information in the form of a sentence that reads similarly to how the reader would phrase the question. For example: "Display contracts with expiry dates more than 30 days ago."

Iteration: Repeat the process

This design can be further improved by repeating the preceding steps of the problem analysis as often as time or diminishing returns permit. For example, an additional iteration might identify the fact that in a typical interface, users must click an "OK" button to instruct the software to begin the search for the specified contracts. In reality, "OK" really means "Display what I've just specified," and it makes sense to match the action produced by clicking a button with the label that describes this action. This suggests yet another simplification: Remove the word "display" from the heading, and place it on the button that currently says "OK." The revised design appears in figure 13.4.

Further iteration, combined with usability testing to ensure that the results are as good as we think they are, could generate additional design improvements. The simple, iterative analysis of the problem produces an impressive decrease in the amount of space required to let users select which contracts they want to see. The resulting design also becomes substantially clearer and more usable than the original, because the wording matches how users would define the problem and the sequence follows that in which users would typically define the question and initiate the action. Raskin (2000) provided a much more detailed discussion of interface efficiency that builds on what I have presented thus far.

Figure 13.4. An improvement on figure 13.3 in which the button that launches the activity clearly describes the activity performed.

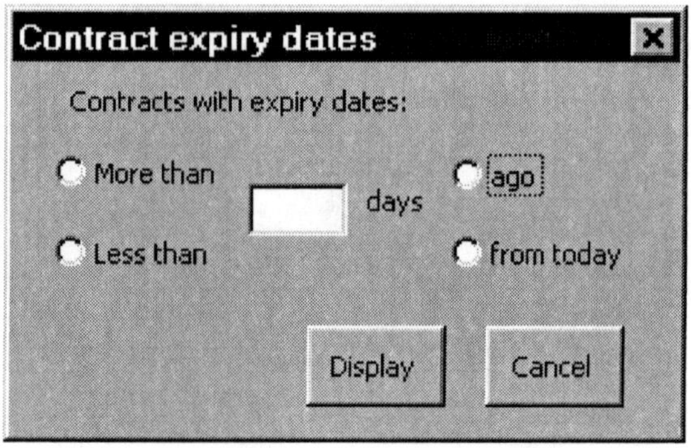

Example 2: Improving an actual design

My second example involves a task-management system that my colleagues and I developed at work as part of an exercise in which our goal was to speed up the production of our technical reports. The system, based on Microsoft Outlook's "task" functions, helps us to assign and manage responsibilities and deadlines. The original dialog box produced by the developers (figure 13.5) exhibited several of the problems typical of first-draft designs, including:

- Different categories of information were mixed in a single display area rather than being separated by type of information: for example, the actual task (labeled "TO DO:") is concealed amidst the title and other information that identify the report.
- Many labels were unclear: Discussions with users of the form confirmed our initial suspicions that the button labeled "Save_exit" was incomprehensible, since Outlook's default "Save and close" button at the top left of the screen seemed to do exactly the same thing—but did not.
- Other labels were misleading or unhelpful: The heading "REPORT TASK MANAGEMENT" actually introduces material that identifies the report, and does not actually communicate any information that users would consider to be "management."
- Many labels were redundant: The heading "REPORT TASK MANAGEMENT" is unnecessary, because the user cannot access this screen without consciously choosing to work within the report task management system and since the heading

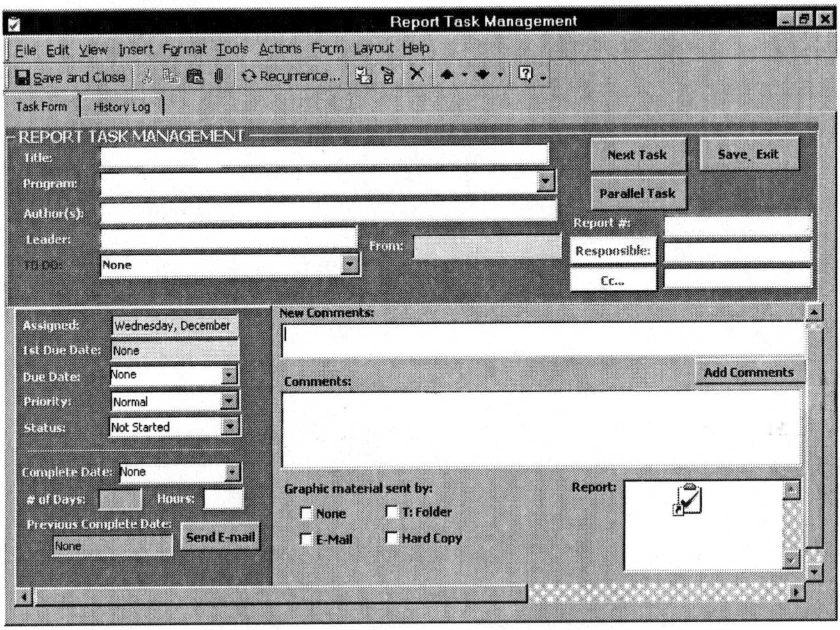

Figure 13.5. The original dialog box designed by the developers.

already forms the title of the dialog box. The redundancy provides little benefit, and requires users to read and filter out more text than should be required.

- There was needless repetition of information, yet useful redundancy was missing: The word "date" appears no less than four times at the left of the screen, yet is not present beside the field "Assigned:" to explain the relevance of that field.
- Visual hierarchies were not respected: The entire section at the left, which begins with the field "Assigned:", lacks a heading and is separated by a horizontal line into what appear to be (but are not) two functionally different areas.
- Labeling of fields was visually inconsistent: Typographic cues were poorly chosen. Moreover, fields such as the report title and program, which are entered when the task is first created and cannot be edited from within this dialog box, are displayed with the same white background used to denote editable fields such as the two Comment fields.
- The sequence did not reflect the user's expected workflow: Buttons such as "Next task" that appear at the top of the dialog box, suggesting they can be clicked before filling out the rest of the form, should instead appear at the bottom, where the user reaches them only after reading and filling in the required fields.

Redesign based on the process described for example 1

Identifying and resolving these problems let us produce a proposal for redesign, which produced the revised dialog box design in figure 13.6. Although the developers did not make all the changes we recommended, they did accept many of our recommendations, including the following:

- Different categories of information are now grouped in their own display areas: For example, the information that identifies a report now appears in its own section of the screen, under the heading "Info." Similarly, the task information now occupies its own distinct grouping to the right of the identifying information.
- Labels are now substantially clearer: The "Save_exit" button has been relabeled "Save & send," and a popup tooltip that appears when the users holds the mouse cursor over the button explains its use. Whereas "Save and close" saves changes to the status of the task, as in all other Outlook tasks, the new button is used only once a task is complete, and lets you define a new task and send it to someone else. (In a subsequent design phase, this button should be greyed-out until the current task is complete to indicate that the button is unavailable.)
- Labels now correctly identify the information they introduce: For example, the dates and times now fall under the single heading "Dates Time." (In a subsequent design phase, this label will probably become "Dates and times" or "Schedule.")
- Redundant labels have been removed or moved to more relevant positions: For example, the heading "REPORT TASK MANAGEMENT" now appears on a tab that defines the portion of the available information that currently appears on the screen; this labeling and the use of tabs sends the message that "this part of the dialog box relates to task management; the next tab displays the workflow diagram; the next...." (In a future revision, the label should no longer use entirely capital letters, thereby improving legibility.)
- Repetition of information has been minimized by including shared information in headings, while simultaneously restoring useful labels to information that formerly lacked them: For example, the label "Dates Time" now appears only once, as a heading that introduces *all* date-related information.
- The new design respects visual hierarchies to a much higher degree: In addition to more consistent and legible typography, the discrete sections of the dialog box have been grouped

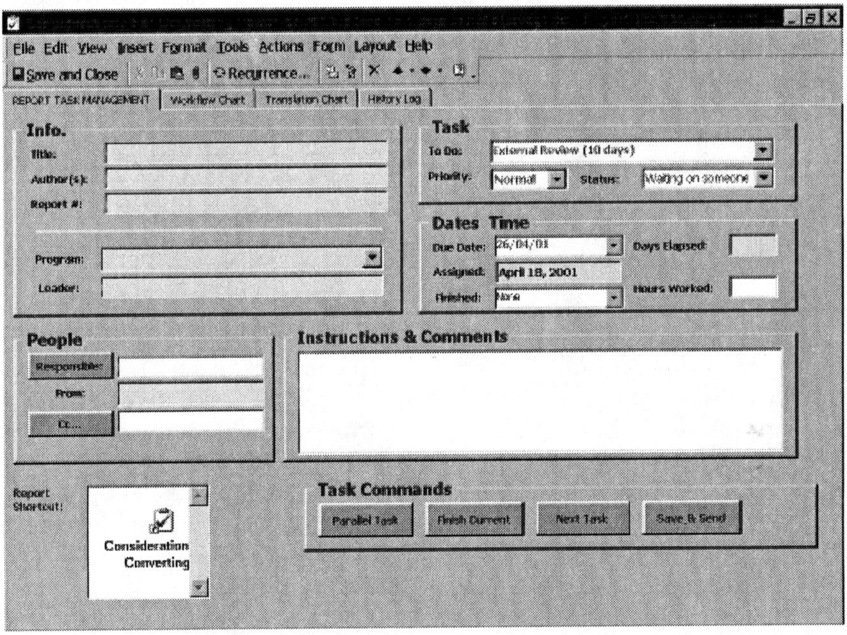

Figure 13.6. A redesigned dialog box based on information-design principles.

into visually homogeneous areas using bounding lines and drop shadows.

- Labeling of fields is now visually consistent: All uneditable fields, such as those that display the report title and program, now have a grey background, whereas all editable fields have a white background.

- The sequence of information now reflects the user's expected workflow: Moving from top left to bottom right, the standard reading sequence in English, readers first see contextual information that identifies the report, additional context (what must be done with that report), and deadlines (dates and times). Information that is redundant to the person assigned the task (gathered under the "People" heading") but relevant to managers who must review the status of a task appears next, followed by room to display any unique instructions that relate to this particular task. Just as in the top-down hierarchy used by Racine and Crandall (2001), matching the sequence to the user's needs increases usability.

Additional changes were made based on consultation with users of the report task management system and the application of various information design principles. These include:

- Creating a single, standard directory for each report, accessible by all users on the network, made the entire portion of the screen in Figure 13.5 labeled "Graphic material sent by" unnecessary. Graphics now reside at a predictable location on the network.
- Tabs labeled "Workflow" and "Translation" were added to present flowcharts of the overall sequence of tasks required to produce a report; in effect, this serves as online help for part of the dialog box. Moreover, to remind users of this task sequence without requiring them to leave the dialog box and invoke the Windows help software in a new window, the popup menu under the "Task" heading contains a complete list of all possible tasks, in the order in which they must be performed. This type of affordance has many of the advantages of "embedded help" (Ray & Ray 2001); in addition, it integrates assistance (defining the task order) with the application so tightly that users may never realize that they're using a help system. By knowing the task they just completed, users can simply scroll one line further down the menu to select the task they will assign to the next user. This design removes the need to memorize the task order and minimizes the need to consult the workflow diagrams.
- The comments and instructions were combined into a single editable field, with no requirement to click a separate "Add comment" button (as was the case in the original design in figure 13.5). Because no other portion of the dialog box required a button to implement changes, this greatly improved the consistency of the interface. Using two separate fields for comments had originally been deemed necessary to prevent users from inadvertently altering or deleting previous comments, but subsequent analysis revealed that these comments could simply be added to a log of all comments and all actions related to a task, accessible via the "History Log" tab.
- Color was used more judiciously to label fields: For example, the due date appears in red to make it stand out from the other dates.

Iteration: Repeat the process

The design that appears in Figure 13.6 represents a work in progress, as the improved interface had not yet been reviewed and revised at the time this chapter was written.

Additional problems remain to be resolved, such as:

- Unnecessary labels must still be removed: For example, the heading "Dates Time" may eventually split into two sections: "Dates," to present date information, and "Times" to present time information. Alternatively, it may be relabeled "Schedule."
- Although the four buttons appear on every screen, some buttons are unusable at certain times or for certain tasks: For example, the "Next Task" button cannot be used before the current task is complete, yet it remains highlighted, suggesting that it is an available option. In a future revision, buttons should be dimmed when not available to avoid misleading users.
- Some information relevant to managers (e.g., the program number under which the report is being produced and the program leader) is not relevant to users of the task system, for whom the title and author of the report suffice to identify the report. Similarly, the "Days elapsed" field, which the software calculates automatically, is irrelevant to most users. Moreover, the value for the "Hours worked" field must be estimated rather than calculated automatically, leading to imprecisions. Placing the less relevant information on a "Management statistics" tab would further simplify the screen.

Conclusion

Interface design involves many intangible factors, and because interfaces often encapsulate complex information about information and the tasks users perform with that information, interface design often seems to be a work of art rather than science. Nonetheless, a few simple techniques can let designers rigorously examine a design to identify means of improving it. This examination process involves grouping of related information, elimination of unhelpful redundancies, and the development of new affordances, and continues with an iterative review of the resulting design by repeating these three methods to determine whether further improvement is possible.

Soliciting feedback from the users of the interface can identify additional problems and offer further opportunities for improvement, but also provides an important reality check that the design truly does represent an improvement. However, technical communicators often complain that their company restricts access to actual users of the software they're documenting. In the absence of an opportunity to test an interface with real users, Raskin (2000) provided considerable food for thought about how to analyse the *theoretical* efficiency of improved interfaces.

References

Bertin, J. (1983). *Semiology of graphics*. Madison: University of Wisconsin Press.

Coe, M. (1996). *Human factors for technical communicators*. New York: Wiley.

Cooper, G. (1998). Research into cognitive load theory and instructional design at UNSW. School of Education Studies, University of New South Wales, Sydney, Australia. www.arts.unsw.edu.au/education/CLT_NET_Aug_97.html.

Horton, W. (1994). *Designing and writing online documentation*. New York: Wiley.

Mandersloot, W. G. B. (1996). Logical criteria applied in writing and in editing by text analysis. *Journal of Technical Writing and Communication, 26*(1), 57-68.

Moore, P., & Fitz, C. (1993). Gestalt theory and instructional design. *Journal of Technical Writing and Communication, 23*(2), 137-157.

Racine, S. J., & Crandall, I. B. (2001). Retrieving product documentation online. *Technical Communication, 48*(1), 31-41.

Raskin, J. (2000). *The humane interface: New directions for designing interactive systems*. New York: Addison-Wesley.

Ray, D. S., & Ray, E. J. (2001). Embedded help: background and applications for technical communicators. *Technical Communication, 48*(1), 105-115.

Contributors

Michael J. Albers is an assistant professor at The University of Memphis, where he teaches in the professional writing program. He completed his PhD in Technical Communication and Rhetoric from Texas Tech University. Before returning to work on his PhD, he worked for 10 years as a technical communicator writing software documentation and performing interface design. His research interests include the history of information design in early books, information design focused on answering real-world questions and online presentation of complex information.

Peter Blijleven is a research assistant at the Faculty of Educational Science & Technology of the University of Twente in Enschede, the Netherlands. He carries out his research activities mainly within the field of curriculum-technology, multimedia-learning and teacher education.

Saul Carliner focuses on content design and development for the workplace. He has advised Berlitz, Guidant Corporation, IBM, Microsoft, and state and federal agencies, among others. He has served on the faculty of Bentley College and the University of Minnesota, and as visiting faculty at the City University of Hong Kong and the Pan African Institute for Development. He is the author of *Designing E-Learning, An Overview of Online Learning,* and the Outstanding articles of 2000 and 2001 in *Technical Communication.* Saul is a Fellow and past international president of the Society for Technical Communication (STC). He holds a PhD in instructional technology from Georgia State University.

Maribeth C. Clarke is a teacher educator and associate professor in the School of Family Life at Brigham Young University. Her PhD from Utah State University is in Psychology—Research and Evaluation Methodology with an emphasis in mailed questionnaire survey response rate methodology. She is a family life education generalist who uses her methodological research expertise to network with specialists on family issues both in and out of her discipline.

351

John Duda is a vice president at Bank of America and manages the user experience for corporate and commercial products. He is responsible for the information architecture, user interface and usability testing of financial web-based products. John's experience includes over ten years of UI and interaction design for various corporations, as well as assisting in establishing the Usability Testing and Evaluation Center at IIT. John received his M.S. in Technical Communications and Information Design from IIT.

Susan Feinberg is a Fellow of STC, Professor of English and Director of the Usability Testing and Evaluation Center at Illinois Institute of Technology. The goal of this Center is to promote usability testing as part of the product development cycle, especially for web-based instructional materials. Learning theory as it pertains to web-based and other products is a major focus of the research performed in the Center, research that can contribute to a greater understanding of how people learn.

Geoff Hart is an editor, writer, and French translator with more than 15 years of experience in the fields of information design and technology transfer. Although his background has primarily been in communicating the results of scientific research to diverse audiences, an increasing proportion of his work involves assisting in the design of software user interfaces and designing the online help systems that support their use. Geoff serves on the editorial advisory board for *Technical Communication*.

Leanne Jansen is an adviser at CPS Educational Development and Advice in Amersfoort, the Netherlands. CPS offers advice, carries out research, and develops materials for the educational field.She worked formerly at the Faculty of Educational Science & Technology of the University of Twente.

Beth Mazur is the founder of STC's Information Design special interest group (ID SIG). She is the editor of the SIG newsletter, *Design Matters*, and author of the ID SIG's weblog, IDblog. She has more than 20 years of experience in technology and communications, working most recently as a web developer. She is a doctoral student in communication design at the University of Baltimore and has a BA in Mathematics from Boston University and an MA in Communication, Culture, and Technology from Georgetown University.

Barbara Mirel has studied usability and complex problem solving for the past 15 years as both an associate professor and human factors manager in industry and healthcare. She currently is teaching in the School of

Information at the University of Michigan and consulting independently on user research and usability testing. She has published widely in technical communications and has served as a book editor and article editor for two journals in the field. She is a recipient of the ACM-SIGDOC Rigo award and is an Associate Fellow in STC.

Margaret Murphy is Communications Coordinator for the Department of Chemical and Environmental Engineering at Illinois Institute of Technology. She is responsible for image development and communications with all internal and external audiences via print and electronic media. She also serves as facilitator between faculty and distance education staff in the development of the department's internet curriculum. Ms. Murphy holds an M.S. degree in Technical Communications and Information Design from IIT.

Whitney Quesenbery is a user interface designer and helps companies implement a user experience process to improve the usability of their products. As a principal at Cognetics, her projects ranged from hypertext to hospital management software and included work developing the LUCID (Logical User Centered Interaction Design) Framework. Whitney is active in professional societies as a manager of the STC Usability SIG and a member of the UPA Board of Directors. She has a degree from Bryn Mawr College.

Ann Rockley is President of The Rockley Group, Inc. Ann has an international reputation in the field of e-content, enterprise content management, and e-learning. She has 20 years experience developing all types of electronic information ranging from Help to sophisticated customized electronic content management systems. Ann teaches Information Design for Multiple Media and Enterprise Content Management at the University of Toronto. She has a Master of Information Science degree from the University of Toronto, and is an Associate Fellow of The Society for Technical Communication.

Karen Schriver is an international leader in information design. She is the former codirector of Carnegie Mellon University's graduate program in professional communication. She also held the Belle van Zuylen Professorship at the University of Utrecht in the Netherlands and a visiting position at the University of Washington in Seattle. Karen now heads her own company, KSA, a firm that helps organizations improve the quality of their communications. Karen is the author of *Dynamics in Document Design: Creating Texts for Readers* (Wiley & Sons). She is now working on a book about the nature of expertise in information design.

David Sless is Director of the Communication Research Institute of Australia, Professor in Science Communication at the Australian National University, and Senior Research Fellow at the Information Design Research Centre at Coventry University. In 2001, he became Co-Chairman of the Information Design Association in the UK. David is a Fellow of the Royal Society of Arts. He graduated from Leeds University in 1965.

Karl L. Smart teaches business communication and document design courses in the Department of Business Information Systems at Central Michigan University. His current research focuses on applying quality and information design to the Internet. Prior to his academic appointment, he spent several years working in the software industry. He is active in several professional organizations and is a contributing editor for STC's magazine, *Intercom*.

Hans van der Meij works at the Faculty of Educational Science & Technology of the University of Twente in Enschede, the Netherlands. His research interests are questioning, instructional design, minimalism, and usability testing. He has written 98 chapters and articles and received numerous awards.

Jean Vanderdonckt holds a Ms. both in Mathematics and Computer Science and a Ph.D. in Computer Science, all from University of Notre-Dame de la Paix. He is currently professor at the School of Management (IAG) of Université catholique de Louvain (UCL), Belgium,. He is leading the Belgian Lab. of Human-Computer Interaction..

Beverly B. Zimmerman is Associate Professor of English at Brigham Young University where she teaches technical communication and oversees the English Department's internship program. In 1998, she received an STC research grant to study the effectiveness of information design principles in creating computer-generated health questionnaires. She holds a PhD in Instructional Design.

Author Index

Subject Index